The Underwater Eye

With warm wishes —

Margout

ps I want to know about
your family's history with
underwater cameras!

The Underwater Eye

HOW THE MOVIE CAMERA OPENED
THE DEPTHS AND UNLEASHED
NEW REALMS OF FANTASY

MARGARET COHEN

PRINCETON UNIVERSITY PRESS
PRINCETON & OXFORD

Copyright © 2022 by Princeton University Press

Princeton University Press is committed to the protection of copyright and the intellectual property our authors entrust to us. Copyright promotes the progress and integrity of knowledge. Thank you for supporting free speech and the global exchange of ideas by purchasing an authorized edition of this book. If you wish to reproduce or distribute any part of it in any form, please obtain permission.

Requests for permission to reproduce material from this work should be sent to permissions@press.princeton.edu

Published by Princeton University Press
41 William Street, Princeton, New Jersey 08540
6 Oxford Street, Woodstock, Oxfordshire OX20 1TR

press.princeton.edu

All Rights Reserved
ISBN 9780691197975
ISBN (e-book) 9780691225524
Library of Congress Control Number: 2021948741

British Library Cataloging-in-Publication Data is available

Editorial: Anne Savarese and James Collier
Production Editorial: Karen Carter
Jacket/Cover Design: Pamela Schnitter
Production: Erin Suydam
Publicity: Alyssa Sanford and Amy Stewart
Copy editor: Beth Gianfagna

Jacket/Cover Credit: Film still from *Below the Sea* (1933), directed by Albert S. Rogell. (Background) koko-stock / Deviant Art

This book has been composed in Arno

Printed on acid-free paper. ∞

Printed in the United States of America

10 9 8 7 6 5 4 3 2 1

In memory of my parents, who shared their love of vision.

"a drawing room in the depths of a lake . . ."

—ARTHUR RIMBAUD

CONTENTS

ILLUSTRATIONS

Introduction

Part 1: Vision Immersed

Part 2: The Wet Camera

Part 3: Liquid Fantasies

Epilogue

Color plates for figures 0.3, 1.1, 1.3, 1.4, 1.5, 1.6, 1.7, 1.26, 2.2, 2.10, 2.12, 2.14, 2.15, 2.16, 2.19, 2.21, 3.3, 3.17, 3.22, 3.28, 3.31, 3.36, 3.37, 3.38, 3.40, 4.2, 4.3, and 4.4 follow page 124.

NOTE ON TIMESTAMPS

In order to help the reader locate media sequences discussed, I have included timestamps throughout the text, notes, and figures. I preface these timestamps with the abbreviation "ca." (circa) because the digital time counts given for the location of specific frames may be imprecise. Different copies exist of the same moving image works, with some different opening sequences, and also occasionally edited differently. Such variation can alter the time count— sometimes just a second or two, but occasionally up to a minute and a half.

ACKNOWLEDGMENTS

I am grateful to Stanford University, whose Deans of Arts and Sciences have supported this book across a decade of research and writing. I began the project as a Violet Andrews Whittier Faculty Fellow at the Stanford Humanities Center in 2011–12, when I was supported as well by the Stanford Humanities Center. I pursued research while leading a Stanford Humanities Center and Stanford Arts Institute collaborative grant in 2013–15, "Seeing through Water," including a joint project with a team from the Sydney Environment Institute led by Iain McCalman and a team from Vanderbilt University led by Jonathan Lamb. In 2016–17, I received a Humanities and Arts Enhanced Sabbatical Award from the Stanford Dean of Arts and Sciences. I thank the John Simon Guggenheim Memorial Foundation, which named me a fellow in 2017–18, providing a research year supported by the Stanford Dean of Arts and Sciences as well.

Valérie Pisani, curator of the Collections artistiques at the Musée Océanographique de Monaco, shared the treasures of its library and art collection. Peter Brueggeman, retired Head of Scripps Institution of Oceanography Library and Archives at the University of California, San Diego guided me through the collection's extraordinary resources in 2012–13 and answered subsequent queries as the project took shape. He introduced me to Elizabeth Shor, who shared her archival research on Pritchard, and to Dale Stokes, who provided insight into the intersection of undersea art and scientific diving. I am grateful to Michael Jüng, Director of the Hans Hass Institut für Submarine Forschung und Tauchtechnik for sharing archival material, as well as to Reg Vallintine of the Historical Diving Society for a copy of *Diving to Adventure*. Joe Wible and then Donald Kohrs have kindly assisted my use of resources and rare books at the Miller Library of Stanford's Hopkins Marine Station.

I thank Victoria Googasian for her fine research on the popular reception of films discussed; Matthew Henry Redmond, who reviewed the manuscript with care; and Andrew Tan for help with the imagery and for his precise and thoughtful fact-checking. At Princeton University Press, I am fortunate to have as editor Anne Savarese, who has guided me in refining the book with keen insight and great common sense. I thank James Collier for solving myriad problems across production and Beth Gianfagna for her attentive copyediting and responsiveness throughout the process.

The Underwater Eye developed from my initial curiosity about the historical context surrounding Jules Verne's *Twenty Thousand Leagues under the Seas* into a study of underwater film thanks to discussion with colleagues in fields ranging from nineteenth-century studies to film studies to the environmental and notably oceanic humanities. I have learned unceasingly from their questions in conferences, lectures, and informal conversations, as well as from their responses to my writing on the subject, from fellowship proposals to first drafts for a writing group to feedback on articles. It is not possible to express concisely my appreciation for so much rich dialogue over so many years: the vibrant intellectual exchange of our community gives meaning to research and academic life.

To my close friends and family, my thanks go beyond words. Anne, let's just keep swimming. Bill, how did I get so lucky? Dearest Max and Sam, you've sustained me all along this book's writing and no time more than over the past difficult year.

The Underwater Eye

Introduction

Of all the environments on our planet, the underwater realm is the most remote. Across history, societies that have lived in contact with the sea have gleaned information about the depths from surface observation, from what they could fish, and from what washes up on shore. Free divers, too, have brought back knowledge, although they have been severely limited by an atmosphere that is toxic to humans. In water, we cannot breathe, our eyes have trouble focusing, and as we descend, the pressure on the body becomes unbearable. Without technology, humans could not spend sustained time below. Since antiquity, people descended in diving bells for harvesting, salvage, and warfare. The premodern diving bell, however, was extremely dangerous and did not offer much of a window into the depths, as it was immobile and oftentimes used in murky waters.[1]

Western technologies made sustained underwater access possible in the middle decades of the nineteenth century, when the Industrial Revolution at sea created both the demand and the inventions for diving. The modern diving suit permitting prolonged submersion utilized steam and pistons, which had the power to force the lighter atmosphere of air into the denser atmosphere of water. Across the late eighteenth and early nineteenth centuries, engineers experimented with ways that industrial technologies might further human presence below. One among many experiments was the *machine hydrostatergatique*, designed by a man named Fréminet, as Jacques-Yves Cousteau explained in the film *The Silent World* (1956), standing before a 1784 print showing this invention in the cabin of his ship (fig. o.1).[2] Cousteau chose to highlight Fréminet's machine among the experiments of this era because it included a separate chamber that held a reservoir of air and was thus the forerunner of scuba. A decisive innovation in the history of diving was the closed helmet bolted to the body of the suit, such as the design manufactured by German-English

FIG. 0.1. Jacques-Yves Cousteau and Louis Malle, *The Silent World* (1956), ca. 5:14. The match flame reflects off a 1784 print showing Fréminet's *machine hydrostatergatique*, a forerunner of scuba, and thus a creative spark intimating Cousteau's invention.

Augustus Siebe in 1839, which kept the diver's head encased in air, even amidst underwater turbulence (fig. 0.2). With the assistance of designs such as Siebe's, humans could reliably breathe underwater, and they also could see through the helmet's glass windows. Finally, they were able to spend extended periods in a realm that had for millennia been hidden from view.

Closed-helmet diving suits revealed that this environment was dramatically different from land, starting with its physical conditions. As anyone today who puts on goggles to go for a swim can observe, the aquatic atmosphere is cloudy instead of transparent, and when we look into the distance, particularly when the light is not directly overhead, objects rapidly fade into a fog. Further, we perceive colors differently; even at a few feet below the surface, reds are muted. Water is eight hundred times denser than air, and it absorbs light waves so effectively that even in crystalline water the furthest the human eye can see is around 240 feet.

And yet, when industrial technologies first enabled access to the underwater world, there was startlingly little public or scientific interest in firsthand observation, although the depths of the seas span almost three-quarters of the

FIG. 0.2. Augustus Siebe, "First Closed Diving Helmet and Breastplate" (1839). Photograph © Science Museum/Science & Society Picture Library.

globe and sustain terrestrial life.[3] While readers in the early modern and Enlightenment eras avidly consumed accounts of overseas voyages, nineteenth-century interest in the diving suit and the windowed diving chamber initially remained confined to industry, to serve salvage and construction operations for the ever-growing boom in shipping and travel. Scientists seemed indifferent to diving, with the exception of a few naturalists such as Henri Milne-Edwards, who described an expedition to Sicily in the 1840s that included scientific diving. Scientific indifference is all the more remarkable because diving technologies initially were deployed in shallow coastal waters, important arenas of observation for marine biologists at the time.

In contrast, marine scientists took note of engineering discoveries about the deep sea, which, by the 1850s, began to attract public and professional attention as a new realm for observation and exploitation.[4] The challenging project to lay a submarine communications cable across the Atlantic, launched in 1854 by American entrepreneur Cyrus Field, was particularly influential, and it eventually succeeded with the collaboration of British and American engineers and governments. People had long believed the seafloor to be an even surface, since mariners for centuries had no need to take its soundings (ships are safe in deep water). Underwater mountains and valleys, however, hitherto unknown, presented formidable obstacles for running a cable, and the project was not completed until 1866. As engineers worked on laying the cable, they fished up unknown creatures, showing biologists that depths previously thought sterile held unanticipated life. These developments confirmed the unprecedented, alien qualities of the undersea and promoted the expansion of what historian Rosalind Williams has called "human empire" into the depths.[5] Great Britain was one of several sea-oriented nations to launch pioneering scientific expeditions, such as the four-year global voyage of HMS *Challenger*, from 1872 to 1876. Nonetheless, scientists on this voyage made their observations from the ship's deck, although it would have been conceivable to use diving suits near shore.

Nor was helmet diving of interest to amateurs, despite the growing appeal across the nineteenth century of the seacoast as a leisure destination. What historian Alain Corbin notes as "the lure of the sea" emerged among upper classes in the later eighteenth century, who began to view the sea as promoting health and soul-stimulating pleasures, in contrast to their previous aversion.[6] Across the nineteenth century, the lure of the sea grew more democratic, as railways facilitated coastal access, creating demands met by the construction of new resorts. Yet such trips did not include opportunities to view or experience submarine engineering. Diving was a labor-intensive activity that required considerable resources as well as heavy and cumbersome suits. Even so, it is not impossible that people might have been afforded the chance to try a dive helmet. If there had been interest, perhaps there might have developed an early version of today's snuba, a less technical and more accessible form of leisure diving than scuba, where swimmers dive in shallow depths tethered to an air tank at the surface.

While scientists and members of the general public did not take observation down into the sea, they were eager to have the sea come to them. Aquariums were first developed to enable biologists to observe living marine creatures. Victorian naturalist Philip Henry Gosse oversaw the design and installation of the first public aquarium, the London Zoo's "Fish House," which opened its doors to the public in May 1853. Through its large glass-sided tanks, audiences enjoyed vistas of marine creatures in their environment, revealing the vibrant beauty, strange forms, and fascinating movement of even such humble invertebrates as sea anemones and jellyfish. Public installations proliferated in zoos and in pop-up locations, from world's fairs to fisheries exhibitions. Further, people were keen to keep aquariums in their homes. A year after the London Zoo's "Fish House" opened, Gosse published *The Aquarium: An Unveiling of the Wonders of the Deep Sea* (1854), combining information on how to stock and maintain a home aquarium with details about how to collect specimens on trips to the coast. Gosse's book helped launch a type of popular guide promoting amateur marine naturalism, encouraging children and adults to collect coastal specimens and to keep aquariums for domestic enjoyment.[7]

Now, finally, people were able to access the "bejewelled palaces which old Neptune has so long kept reluctantly under lock and key," a reviewer of Gosse's *The Aquarium* declared.[8] Yet, public enthusiasm notwithstanding, aquariums were artfully curated gardens, underwater installations created to share scientific knowledge and also to delight. Their tanks were not facsimiles of the depths, and, moreover, their curators did not replicate real submarine conditions. Such knowledge would not have been hard to obtain; it was just a little under two and a half miles from the London Zoo in Regent's Park to 5 Denmark Street, London, the headquarters of Augustus Siebe's firm, recognized as a leader in industry diving.

Aquariums are commonly acknowledged as the first media form to give general audiences access to the underwater realm. So associated was the aquarium with the undersea that Jules Verne used it as a model in *Twenty Thousand Leagues under the Seas* (1870), the first book to imagine the planet as spanned by a holistic yet diverse undersea environment. A popularizer of science and technology, Verne wrote the novel soon after the completion of the transatlantic cable project. He consulted contemporary oceanographers and biologists, also basing his fantasies on nineteenth-century diving inventions. Yet he did not go to working divers for his submarine descriptions, which transposed terrestrial optics beneath the surface. For example, on the captivating hunting expedition in the forest of Crespo, thirty feet beneath the surface of the ocean, Verne's professor narrator Aronnax observes, "The sun's rays struck the surface of the waves at an oblique angle, and the light was decomposed by the refraction as if passing through a prism. It fell on the flowers, rocks, plantlets, shells and polyps, and shaded their edges with all the colours of the solar spectrum. It was a marvel, a feast for the eyes."[9] In fact, the aquatic atmosphere is hazy;

nor does it decompose light frequencies as a prism does. Further, the spectrum of color humans perceive in the depths diminishes, starting with the disappearance of red at about fifteen to twenty feet.[10]

Why were general audiences, to say nothing of scientists, indifferent to penetrating this vast new environment when immersive technologies appeared? We can imagine governments and entrepreneurs expending resources for exploration, had there been sufficient public interest. The Arctic Circle still stubbornly retained the lucrative promise that navigators could chart a Northwest Passage to the East, despite centuries of failure, and the doomed John Franklin expedition departed from England in May 1845. Historian of science, technology, and media Natascha Adamowsky suggests a primal emotional reason: the abiding fear of the depths.[11] Perhaps social stratification was an issue as well. The dangerous activity of helmet diving was carried out by working-class divers, while managers typically remained on the surface, in contrast to overseas expeditions, where the literate officer class participated firsthand in the rigors of maritime travel and penned eloquent accounts.

While I can only speculate about absence, it is possible to pinpoint when general publics, amateurs, and scientists alike became excited to learn about the depths: the moment when film was able to capture the submarine environment for audiences on land. In 1912, American J. E. Williamson invented the "photosphere," a submersible that enabled underwater filming, with the operator and camera enclosed in a chamber of air.[12] In 1914, Williamson assembled footage taken from the photosphere and released the first underwater documentary, shared initially with scientist audiences and then distributed more widely, under different titles (*The Terrors of the Deep, Thirty Leagues under the Sea*). Although most of the film is now lost, records of its impact remain, and other photosphere films survive. In the words of Charles J. Hite, president of the Thanhouser film corporation, cited in a 1914 article from the *Moving Picture World*, "No man, until the Williamson invention was made practicable, could tell of the life below. The new invention brings to science the sea's actualities of life, the long lost ships, the Imperators of other days, the hidden reefs, the variegated corals, the moving things."[13]

The Williamson photosphere inaugurated a new era in curiosity and consciousness about the undersea. Submarine film had been preceded by a reliable process for submarine photography developed in the 1890s by French marine biologist Louis Boutan. Yet only with Williamson's invention did people grasp that this technology brought new areas of the planet into view.

Such was the prevailing credit given to film that when the *Photographic Journal of America*, the oldest photography review in the United States, surveyed the brief history of underwater photography in 1921, it opined: "The most satisfactory results have been obtained by the Williamson camera, the results of which are familiar over the world by their incorporation into movie shows."[14] Another example of the authority accorded submarine film was the source of the underwater photographs featured in the first full-length popular naturalist book about diving, William Beebe's *Beneath Tropic Seas* (1928), which were movie frames taken from the photosphere.

Film's role in stoking submarine curiosity was in part due to historical coincidence. Film became the visual medium with the greatest documentary authority in the twentieth century when the human ability to experience the underwater realm firsthand was expanding, thanks to dynamic innovations in diving and submersible technologies. At the same time, it is almost as if a muse of history was at work in such coincidence, because with film, the medium fits the message. Moving imagery has the unique capacity to disseminate particularly distinctive, alluring properties of submarine creatures and conditions, such as the glide of movement and the disorienting perception in underwater space.

Print accounts had been the principal medium for describing overseas exploration, and several print narratives would become best sellers for their portrayals of the undersea, once the environment piqued general interest. American William Beebe's 1928 *Beneath Tropic Seas* was widely disseminated, and Jacques-Yves Cousteau's book *The Silent World* (1953), published simultaneously in English and in French, was its equivalent during the postwar launch of scuba. Yet throughout published observations of the undersea, we find the phrase "words cannot express." This rhetorical term, the *je ne sais quoi*, dates to the early modern era to underline the writer's failure to convey an intense experience—that, in twentieth-century dive narratives, was the perception of extraordinary submarine reality.[15] Thus, Beebe wrote at the beginning of *Beneath Tropic Seas*, "you had planned to tell the others all about it, but you suddenly find yourself wordless," describing surfacing from a dive (and referring to himself in the second person).[16]

Human perception undersea posed difficulties for visual as well as verbal expression, and few images of this realm based on firsthand observation exist before photography. One noteworthy example is *Davy Jones's Locker* (1890), the sole undersea scene by the premier British marine painter William Lionel

FIG. 0.3. William Lionel Wyllie, *Davy Jones's Locker* (1890), oil on canvas. Photograph © National Maritime Museum, Greenwich, London. Wyllie has captured the submarine palette and haze, yet the clearly defined detail in the foreground of the scene, such as the octopus entwined with human remains, characterizes seeing through air rather than water.

Wyllie, which Wyllie drafted after improvising a dive helmet out of a biscuit tin and studying marine life in an aquarium (fig. 0.3). Wyllie's remarkable painting of a ship's wreckage shows us the pervasive haze underwater and its muted palette. Fish flit like silver ghosts in this gloom, and an octopus intertwines with a blanched human bone, perhaps a pelvis, lying next to a skull. True to octopus camouflage, the creature's head has taken on the ghastly pallor of the bone, with brown tentacles protruding from holes in the bone that help viewers identify its presence. Yet the octopus in Wyllie's painting remains a sinister emblem of death, lacking the sense of life imbued by its glide and remarkable undulations.[17] Nor could Wyllie—or early black-and-white film, for that matter—depict the octopus's continuously changing hues, which would fascinate viewers later in the century when filmmakers could capture underwater color.

The primacy of moving picture media in fostering our curiosity about the submarine realm continues to this day. The history of underwater film is one of dynamic technical inventions, both belonging to the history of moving image technology (such as the development of color processes suitable for all types of cameras around 1950), and specific to the challenges of working in one of the most inhospitable environments on earth. Across this history, innovations in both diving and filming continue to reveal new aspects and areas of the underwater environment, feeding and stimulating public interest. The twenty-first-century equivalents of Williamson's *The Terrors of the Deep* are the BBC miniseries *The Blue Planet* (2001) and *Blue Planet II* (2017). Further, as I discuss in my epilogue on *Blue Planet II*, film today is able to show, beyond submarine geology and biology, the impact of the Anthropocene on a realm so long thought to be beyond human reach.

The ability of film to let general audiences dive without compressed air or mask has led us to associate underwater film with documentary. From Williamson's first moving pictures, however, audiences also enjoyed these views for their expressive power, even if they had more trouble articulating such appreciation compared with their marvel at submarine reality. The contribution of the underwater environment to the aesthetics of cinema is the subject of *The Underwater Eye*. By the word "eye" in my title, I mean human vision, in the sense of both knowing and seeing, conceptually as well as sensuously disoriented owing to the physical qualities of the aquatic atmosphere. I also use "eye" in the sense of the movie camera that works as a prosthesis, with the capacity to extend the sense of sight in water, as in other environments on our planet.

With vision immersed, filmmakers have been able to exercise their craft on what I call the underwater film set—including both submarine conditions and technologies for their capture. There, they have created novel, charismatic imagery that expands the visual and narrative imagination. Filmmakers' raw materials for such imagery are aquatic life, place, and even the atmosphere. Immersed in water, the camera exhibits strange, beautiful, and sometimes dangerous submarine flora and fauna, such as sharks, octopuses, or coral. It can also make use of the environment's distinctive optics and its effect on movement, including its attenuation of gravity, permitting exploration of three dimensions, like a liquid sky. The underwater film set has further stimulated new types of adventure beneath the surface, inspired by industry and leisure as well as by the dangers of an atmosphere toxic to human physiology.

FIG. 0.4. Jacques-Yves Cousteau and Louis Malle, *The Silent World*, ca. 1:55. "A motion picture studio 165 feet under the sea" (ca. 2:10).

Jacques-Yves Cousteau and Louis Malle set up shop on the underwater film set in the first minutes of *The Silent World* (1956), their award-winning film vaunting the possibilities of the underwater realm for cinema and human curiosity. "This is a motion picture studio 165 feet under the sea," declares the narrator in the English version of the film, designating a group of divers clustering around a reef, flippers fluttering, wielding torpedo-shaped cameras (fig. 0.4). The glare of their strobe lighting dramatically intensifies the brief spark of a match illuminating Cousteau's cigarette in the scene where he names Fréminet's *machine hydrostatergatique* as a precursor to scuba.

Across my history, readers will see filmmakers create underwater imagery that hearkens back to preexisting fantasies. Pre-Romantic and Romantic aesthetics for celebrating nature figure prominently in underwater scenes, as filmmakers shape areas of the planet that defy viewing habits. Filmmakers submerge stories of thrilling adventures in extreme environments familiar from centuries of European traveling. They devise other figures by reworking famous undersea images from tradition. Romantic aesthetics and a famous undersea image meet in Cousteau's portrayal of underwater shipwrecks in his first film shot with scuba, *Sunken Ships (Epaves)* (1943).[18] Shakespeare famously called the effect of the sea on the physical body and human ambition a "sea-change" in *The Tempest*, when the spirit Ariel conjures for Fernando the strange treasure of his drowned father's decomposing corpse: "Full fathom five thy father lies. / Of his bones are coral made. / Those are pearls that were his eyes. / Nothing of him that doth fade / But doth suffer a sea-change / Into something rich and strange."[19] Cousteau took up this transmutation in *Sunken Ships* but transferred sea-change to the detritus of ships, respecting the taboo against exhibiting human remains. Overgrown with marine creatures, sunken ships become eerie living architecture, recalling the Romantic cult of ruins. Furthermore, Cousteau picked out aspects of a ship's hull that resemble a Gothic arch, sketching the first draft of the wreck as a gothic ruin that he would elaborate with Louis Malle in *The Silent World*.

Many influential fantasies conceived on the underwater film set occur in works whose subject is the sea. Filmmakers have imprinted majestic dolphins, killer sharks, and melancholy shipwrecks on the popular imagination as a result of techniques for organizing images studied here. These fantasies have, moreover, become part of how audiences conceive of this realm that most of us are not able to witness firsthand. Nonetheless, despite the importance of sea subjects in the aesthetics of underwater film, I qualify the environments at issue as *underwater* rather than *undersea*. The underwater film set's contribution to cinema extends to life on land and to the creation of mood. The beauty of underwater movement and the underwater haze can be used to create mystery and glamour, as in the swimming sequence of erotic longing in Jean Vigo's *L'Atalante* (1934) (fig. 0.5) or, on a more sensational note, kinky seduction in a number of James Bond films, starting with *Thunderball* (1965), directed by Terence Young. The swimming pool can turn from banal to dystopically surreal, epitomized by the view of suburban reality that Ben, the hero in Mike Nichols's *The Graduate* (1967), reveals from the bottom of his parents' pool (fig. 0.6). Narrative filmmakers have incorporated the drowned into their use of the underwater atmosphere to evoke social dystopia, in contrast to documentary filmmakers, who eschew showing the dead. Famous dystopic shots include the underwater view looking up at the floating corpse that starts Billy Wilder's *Sunset Boulevard* (1950) and the undulating submerged seaweed and hair of the murdered widow discovered intermingled at the midpoint of Charles Laughton's *The Night of the Hunter* (1955).

FIG. 0.5. Jean Vigo, *L'Atalante*, ca. 1:08:20. Longing in the liquid atmosphere: the captain's runaway wife floats toward him as an alluring vision.

FIG. 0.6. Mike Nichols, *The Graduate*, ca. 23:59. The atmosphere of water and the edges of the face mask distort Ben's view of his parents from the bottom of their pool, expressing twisted suburban values.

The Shapes of Submarine Fantasy

The Underwater Eye is organized into three parts that demarcate eras, according to the state of dive and film technologies, which often work in tandem, affording new possibilities for imagery, whose contours are defined as well by their limitations. I start part 1 with firsthand representations of the underwater environment that circulated in print between the invention of modern helmet diving and the inception of underwater film. These representations include verbal descriptions as well as engravings by the Austrian Baron Eugen von Ransonnet-Villez from a diving bell off the coast of Sri Lanka in the 1860s and paintings conceived and sketched underwater by Walter Howlison "Zarh" Pritchard, who first dove with a helmet off Tahiti in 1904. Through analysis of their works, I introduce readers to the qualities of human perception in the depths that would fascinate, frustrate, and inspire filmmakers.

Part 1 then describes distinctive underwater imagery created in the first era of underwater film. From 1914 until the scuba revolution of the 1940s, both camera and operator were predominantly protected behind glass, whether in a pool or from within submersibles. In narrative cinema, underwater imagery accordingly frames its subjects using the format that film scholar Jonathan Crylen has identified as "the cinematic aquarium," designating rectilinear views, like aquarium scenes.[20] Within this format filmmakers portray adventure, such as the combat with real sharks and a mechanical octopus in Stuart Paton's *Twenty Thousand Leagues under the Sea*. I also discuss the fascination of underwater movement, from the aquatic dance of Annette Kellerman to the glide of sharks. Both movement and adventure take an avant-garde turn in Man Ray's surrealist invitation to look with a marine eye. Part 1 concludes with two Hollywood entertainment films that introduce a shift to the portable camera, which, when coupled with scuba, would transform the stories that could be told underwater and the ability to exhibit drama and suspense. Lloyd Bacon's *The Frogmen* (1951) used the portable Aquaflex camera to portray the underappreciated action of World War II underwater demolition teams. On the cusp of the scuba revolution, Robert Webb's *Beneath the 12-Mile Reef* (1953) captured a sponge diver's view of a tropical coral reef with exquisite underwater footage shot by cameraman Till Gabbani.

In the 1940s and early 1950s, views of the cinematic aquarium were replaced with a vastly expanded range of shots and imagery, as the camera, its operators, and actors were free to roam around beneath the surface. The book's second part considers this transformative era, inaugurated by technologies invented either for or coincidental with World War II. Diving aids included swim fins, masks, and scuba, along with film innovations such as the portable camera and strobe lighting.[21] Such innovations created a range of new ways to exhibit the underwater environment, which dive and film pioneers such as Jacques-Yves Cousteau, Louis Malle, and Hans Hass first explored in documentaries.

Part 2 discusses the shots and imagery created by Cousteau, Malle, and Hass that would help shape the aesthetics of cinema underwater, in narrative film as well as documentary. The directorial choices of these image makers had an impact around the world. As Louis Malle commented about making *The Silent World*, "[w]e had to invent the rules—there were no references; it was too new."[22] Part 2 also discusses influential imagery created during an extraordinary decade of innovation in narrative filming. Works featured include the Walt Disney studio's first live-action film, *20,000 Leagues under the Sea* (1954), directed by Richard Fleischer, where, as in documentary, the film's creators were grappling with how to shape this never-before exhibited environment for audience entertainment. "Everything they were doing down there was experimental," said Roy E. Disney, senior Disney executive and Walt Disney's nephew, about making the underwater sequences of this film, which, notably, implemented techniques from proscenium theater to unify diffuse underwater space.[23]

In addition, I discuss the four-season television series *Sea Hunt* (1958–61), produced by Ivan Tors and shot with a portable camera designed by leading underwater cinematographer Lamar Boren, who did most of the camerawork. A major cultural influence in the United States and abroad, *Sea Hunt* imagined new stories of suspense and adventure based on underwater dangers and the range of postwar military, leisure, and commercial activities practiced undersea. Its innovations in environmentally generated plotting would go on to shape subsequent adventure narratives set underwater, as well as science fiction adventures in the toxic atmosphere of outer space.

In Part 3, I take up the aesthetic contribution of underwater settings to narrative cinema in subsequent decades. Starting in the 1960s, the underwater film set became a dramatic, if not everyday, resource of both films and TV shows, as moving image creators developed aesthetic effects from the possibilities that emerged during the revolution in underwater filming of the 1950s. To describe all the imagery created there is a project exceeding my scope, given the quantity and variety of works, which traverse fiction and documentary, as well as film and TV. I have chosen to describe films that create what I call liquid fantasies: imagery whose aesthetic qualities derive from aquatic environments and that takes on a life in the cultural imagination beyond realistic depiction to inspire cultural myths.

Many liquid fantasies in films, from James B. Clark's *Flipper* (1963) to James Cameron's *Titanic* (1997), amplified storylines and new shots first used in the 1950s, sometimes devising new technologies in quest of more extensive access and yet more dramatic effects. Thus, Cousteau and Malle portrayed a sunken ship as a melancholy gothic ruin when they took audiences to survey the wreckage of the SS *Thistlegorm* in *The Silent World*. Filmmaker James Cameron would echo this gothic mood by incorporating the wreckage of the RMS *Titanic* into his eponymous film. Since it lies at twelve thousand feet, in black depths inaccessible to divers, Cameron commissioned a range of technologies to capture footage, including an ROV (remotely operated vehicle) equipped with a video camera, capable of maneuvering into the wreck's narrow passages and withstanding the extreme pressure in such depths.[24] He otherwise never could have captured views of actual wreckage such as "silt streaming through intricate bronze-grill doors" and "a woodwork fireplace with a crab crawling over the hearth," to cite Rebecca Kegan, clarifying documentary footage artfully incorporated into the film's melancholy montage.[25]

A comparison of Cousteau and Malle's portrayal of a wreck site in *The Silent World* and Cameron's exhibition of interiors of the *Titanic* illustrates the importance of technological access in what filmmakers can and cannot portray. At the same time, while the history of technology is an important thread in my account, dive and film technologies "enable" rather than determine representations, to take up a productive word emphasized by Crylen.[26] The views afforded by technologies are starting points that filmmakers then fuse with aesthetics that further their creative aims. Given how few people have visited these environments, it is worth emphasizing that an underwater setting has no necessary aesthetic frame. Filmmakers can create very different emotional effects with the same type of material. While both *The Silent World* and *Titanic* stage a wreck site in the gothic tradition, Steven Spielberg, in contrast, realized a wreck's possibility as a horror house in an isolated but suggestive moment in *Jaws* (when a diving shark biologist discovers a dead fisherman in his capsized boat).

The aquatic atmosphere imbues movement and optics with unique aesthetic properties, which are difficult to simulate on land. For this reason,

I focus on works created in water, although the qualities of the underwater environment do shape simulated imagery in both live action and animation. Pixar's *Finding Nemo* (2003) and *Finding Dory* (2016), for example, evoke the spectrum of undersea colors and the impact of depth on human perception. With the exception of innovative underwater sequences in *20,000 Leagues*, I do not discuss narratives about submersibles, although a number of influential films about submarines have been set undersea, such as Wolfgang Petersen's *Das Boot* (1981) and John McTiernan's *The Hunt for Red October* (1990). I leave these films to the side because their settings primarily show the inside of vessels, with scant aquatic shots that reveal the hazy outlines of hulls in the murk. Indeed, the menace created by underwater obscurity is an important element of submarine fiction's suspense. The combatants cannot see enemy crafts or torpedoes in the obscure marine environment. Listening and sonar hence become essential to survival.

American works feature prominently across my study, owing both to the importance of Hollywood in the film industry and to the long-standing American coordination of innovations in dive technologies with moviemaking, from the invention of the photosphere by Williamson to James Cameron's ROVs. At the same time, my focus is not on the ways in which such films belong to American cultural history. Rather, I am interested in identifying innovative underwater moving imagery celebrated in both the general press and scholarship, as well as by other filmmakers. Some of the films I discuss are European; some depend on international collaboration. Cousteau worked with Massachusetts Institute of Technology engineer Harold Edgerton to design the strobe lighting that enabled the rich color visuals of *The Silent World*. Cameron, a Canadian based in Hollywood, used Russian submersibles to create the lighting system enabling his deep-sea footage for *Titanic*. In 1956, the Austrian Hans Hass became a household name in the English and German-speaking worlds with the BBC's series *Diving to Adventure*.

For reactions in the general press, I draw on American periodicals because of the practical constraints of archival research, but this American point of departure has led me to essential European works in submarine filmography. In the early 1950s, for example, articles in both the *New York Times* and the niche periodical *Skin Diver*, the first American magazine devoted to diving, featured films by Hass, as well as by Cousteau and Malle. My study is not exhaustive, but I hope it nonetheless conveys a sense of the impact of the underwater environment on the history of moving-image artistry.

Environmentally Specific Imagery: Fact and Fiction

Submarine filming, even for narrative films, offers the opportunity to create aesthetic effects from physical qualities specific to the environment. Murky visibility below, for example, confers a sense of mystery that can be ineffable or menacing, depending on context. The atmosphere of water is denser than air, and thus harder for humans to move through, yet it also offers opportunities for grace and glide. Water-dwelling creatures that extract oxygen by swimming cannot be filmed alive in air. The need to shoot underwater for such depictions, even in fantasy films, blurs the documentary/fantasy divide. True, the atmosphere of air is a real element in narrative films made on land, but we take this atmosphere for granted, except when it is enhanced or disturbed—as, for example, in the lurid colors of sunset amidst smoke, or the palpable quality of light filtering through dust. "The last thing a fish would ever notice would be water," anthropologist Ralph Linton is reputed to have observed. While it may sound obvious, the physical qualities of an environment, even those so familiar they become invisible, play a formative, although not determinate, role in shaping its cinematic expression.

So, too, does an environment's familiarity to its audience. The underwater realm is remote. Even today, comparatively few people dive, and the depths beyond about 130 feet are accessible only to professional divers and submersibles. Further, human perception underwater differs drastically from perception in air. As a result, spectators, particularly of novel scenes, have more difficulty filling out unfamiliar aspects using their imagination—as audiences around the world could, say, if shown the Eiffel Tower from below, or with a close-up of its girders. In introducing viewers to an underwater setting, filmmakers hence take care to introduce orienting elements—for example, through organizing scenes according to well-established aesthetics. Thus, as I have discussed, Cousteau, Malle, and Cameron alike use gothic conventions to present an underwater wreck site.[27] In this way, too, the underwater film set complicates the documentary/fantasy divide.

Another factor that leads filmmakers' work across the boundary between fantasy and documentary relates to challenges in underwater capture. One problem filmmakers face in the marine environment, for example, is the opacity of the water's surface. This opacity makes it difficult to create a visual connection between actions in the depths and corresponding actions that are occurring in the atmosphere of air. Cousteau and Malle addressed this challenge creatively when they filmed sharks feeding on a carcass in *The Silent*

World, by drawing on conventions from horror film scenes of killing. These scenes disconnect the perspectives of killer and victim in such a way as to shock and disturb spectators. Cousteau and Malle did not translate precisely: in *The Silent World*, the victim's limited perspective on the action becomes the crew's topside view from the *Calypso* of a feeding frenzy, while the gaze of the killer becomes submarine viewpoints of the sharks, first as they inspect the carcass, and then, as they feed, their savagery evoked with handheld, jerky close-ups. Cutting back and forth between chaos above the surface and revealing glimpses of action below, the directors sensationalized the shark's predation. In *Jaws* (1975), Steven Spielberg took Cousteau and Malle's horror-film solution for crossing the waterline and returned it to its fantasy inspiration. Now, the victim is associated with the surface view, oblivious to the unseen, prowling shark.

As is the case with the gothic wreck, such a horror portrayal of sharks is an aesthetic choice, and not the only way to organize shark imagery. Cousteau and Malle's contemporary Hass exhibited sharks lyrically in his black-and-white *Humans among Sharks* (*Menschen unter Haien*) (1947/48). Remaining below, Hass's camera invites us to admire these apex predators—warily, to be sure—whom he shows swimming majestically, and to leave them in peace. The Spielberg representation of sharks is so powerful that it has terrified swimmers over nearly the past half century, although in fact the number of attacks by great whites on beachgoers every year is minuscule. In recent decades, marine conservation initiatives are increasingly revealing that overfishing threatens many shark species and that these creatures play a key role in the ocean food chain, which would dramatically alter were they extinct.[28] One means pursued to attract public interest in shark conservation has been to erase the *Jaws* icon and replace it with imagery based on better information. The lyrical *Island of the Sharks* (1999), for example, directed by Howard and Michele Hall, places these animals in an abundant, visually mesmerizing ocean ecosystem.[29] In this framing, the Halls take advantage of the vastly expanded resources of IMAX and color film to continue what I will explain as Hass's fundamentally Romantic aesthetic.

There is one more environmentally specific reason why underwater filming complicates the documentary/fiction divide. Work in the depths is exceptionally difficult, leading productions to rely on a small group of specialists: the same camera people, divers, and engineers overlap in factual and narrative film, and in both, they apply their knowledge of the underwater environment and their eye for how to film it. This overlap starts with Paton's use of

Williamson's documentary footage for *Twenty Thousand Leagues* and goes forward to Cameron's fusion of documentary and studio imagery in transporting spectators to the wreck site of the *Titanic*. In the second half of the twentieth century, Hass, already famous for his documentaries, was the camera person for an early Italian narrative film including scuba footage, *Rommel's Treasure* (*Il tesoro di Rommel*) (1955). Lamar Boren, who filmed *Sea Hunt* (1958–61) and the first James Bond films with underwater action, starting with *Thunderball* (1965), also captured documentary footage for *The Magical World of Disney*'s two-part documentary telefilm *The Treasure of San Bosco Reef* (1968). Cameraman Al Giddings would take up Boren's mantle from the 1970s through the 1990s, filming the underwater sequences in some of the later Bond films and directing underwater photography for Cameron's *The Abyss* and *Titanic*. Giddings also did documentary work across his career, from his cinematography for the Canadian short *The Sea* (1972), directed by Bané Jovanovic, to the IMAX short *Galapagos* (1999), which Giddings codirected with David Clark.

The blurred boundary between documentary and fiction in underwater film notwithstanding, some important differences remain. Documentaries notably respect plausibility, showing events that human observers have witnessed, even if filmmakers sometimes must go to some lengths to reproduce them. Further, documentaries respect location—no such obligation constrains fiction. Ivan Tors, an influential producer of underwater film and TV in the era of the scuba revolution, told *TV Guide* in 1959 that "people never notice 'scientific' errors."[30] Tors was speaking about the most successful production of his career, the TV series *Sea Hunt*, whose underwater scenes stitched together imagery from such diverse locations as the Los Angeles oceanarium Marineland of the Pacific; the warm freshwater of Silver Springs, Florida; and the Pacific Ocean off southern California. When a portion of an episode set in the Hudson River was shot off Catalina Island, "a barracuda, strictly a warm-water fish, swam between [star Lloyd] Bridges and cameraman Lamar Boren in the big scene." On the question of what to do with this implausible detail, Tors told *TV Guide* that "fifteen thousand dollars to take out one fish . . . is nonsense."[31]

In an epilogue, I discuss the way documentary filmmakers integrate their commitment to show reality with technical and aesthetic considerations, introducing yet another formative factor shaping the imagery they create on the underwater film set. This epilogue focuses on the documentary as genre, in contrast to my treatment of documentaries in parts 1 and 2, where I discuss their imagery for its expressive power alone. Thus, I do not inquire whether Captain Wallace Caswell Jr., star of *Killers of the Sea* (1937), in fact slaughtered bottlenose dolphins, nor do I analyze how Cousteau and Malle's exhibition of the SS *Thistlegorm* relates to the actual wreck. In the epilogue, I do examine how documentary persuades audiences to accept its exhibition of submarine reality through the example of the immensely popular TV series *Blue Planet II*, first shown on TV in 2017, which mobilized innovative film and dive technologies to capture never-before-recorded, and in some cases never-before-seen, views of marine life. *Blue Planet II* creators organized such views to entertain but also to prompt action to mitigate climate change, and the series famously succeeded on both counts. As I break down favorite scenes, I show how *Blue Planet II* incited audiences to care about remote oceans and submarine reality by hearkening back to familiar fantasy, even when portraying novel subject matter and, notably, to the liquid fantasies discussed in part 3.

The Blue Humanities in the Depths

In showing filmmakers' creative use of the underwater film set, *The Underwater Eye* contributes to a conversation at the intersection of media studies and marine and maritime studies, a thriving interdisciplinary area that Steve Mentz has named the "blue humanities." Until recently, underwater film was not much discussed in either field. For media studies, the submarine setting is a niche environment, appearing infrequently, and techniques on the underwater film set do not obviously seem to contribute to cinema history. For the blue humanities, the surface of the seas has been the predominant arena for social action, and hence the focus of efforts to reintegrate the seas into history. During the past fifteen years, however, the blue humanities has started to realize that the depths belong to history as well.

Below, I provide details on scholars who have taken the lead in drawing attention both to the cultural significance of the undersea and to submarine film, with the goals of acknowledging work that has shaped my thinking, giving readers reference points in an emerging field, and explaining the contribution of my study. (Those readers who are not part of such conversations may want to skip this and go to part 1.)

The turn to the ocean depths in interdisciplinary ocean scholarship occurred around 2005, the year Helen M. Rozwadowski published *Fathoming the Ocean: The Discovery and Exploration of the Deep Sea*. In this study, Rozwadowski drew together science, history, and cultural studies to reveal how during the mid-nineteenth century the deep sea emerged in the West as a new planetary frontier. Rozwadowski has subsequently pursued her inquiry into the twentieth century, noting the emergence of the undersea as a frontier for strategic, scientific, and technological reasons during the World War II and postwar eras.[32] Stacy Alaimo has played an important role not only in drawing attention to the aesthetic interest of the undersea environment but also in showing the different expressive effects created in its different depths. Thus, she reveals the poetics of the abyssal zone, starting with the writings of William Beebe, whose famous exploits included diving half a mile down in the bathysphere, which he designed with Otis Barton.[33] Alaimo has traced this impact forward to both James Cameron's narrative spectacles and to documentary imagery of the depths in an era of environmental awareness, such as photos of undersea pollution revealing the scale and duration of the 2010 British Petroleum oil spill in the Gulf of Mexico.

The blue humanities' undersea research over the past decade has extended to earlier eras as well. Early modern specialists have focused on the imaginative significance of what was then an almost completely unknown area of the planet, beginning with Steve Mentz's influential *At the Bottom of Shakespeare's Ocean* (2009). The depths have further appeared in crossover studies focused on marine animals, such as Philip Hoare's *The Whale: In Search of the Giants of the Sea* (2010) (first published as *Leviathan, or The Whale* [2008]), Sy Montgomery's *The Soul of an Octopus* (2015), and Richard King's *Lobster* (2011).

Nicole Starosielski is an influential media critic and cultural historian whose *The Undersea Network* (2015) traces the networks of submarine fiber-optic cables around the planet that we take for granted and that enable modern communications technology. Along with her book revealing communication in the hidden depths, Starosielski has written about the role of film in their exhibition, and indeed was one of the first to organize films made undersea into a corpus. She categorizes submarine films according to their ideology of ocean depiction: "exposing the seafloor" (1914–32), "ocean exploitation" (1945–58), and "domesticating inner space" (1960–72).[34] Starosielski points out that although "[a] number of works . . . have begun to address the cultural representations of marine mammals," such as those by Gregg Mitman and Derek Bousé, among others, "these lines of inquiry have yet to develop into a sustained conversation on the cinematic construction of underwater environments."[35] The investigations of Franziska Torma also are in the forefront of expanding understanding of the social and cultural forces shaping undersea film. Torma focuses on the way films are shaped by the conjunction of science, technology, and ideology across the twentieth century and into our present. Thus, for example, she notes the importance of scuba in freeing the diver as pioneering "merman" in the 1950s, while IMAX enhanced the grandeur of blue planet exhibition by the century's end.[36] Building on such studies and also diverging from their primary focus on cultural history, *The Underwater Eye* highlights the contributions of underwater film to the poetics of the moving image.

Inquiries by Starosielski, Torma, and Rozwadowski, notably, are at the inception of a robust conversation around underwater film that has started to develop over the past six years, led by the blue humanities. Crylen is writing a book from his dissertation, "The Cinematic Aquarium: A History of Undersea Film" (2015), which brings technology into dialogue with environmental issues as well as with aesthetics. Natascha Adamowsky notes the continuation of a tradition mixing objective knowledge and wonder, connecting Enlightenment and nineteenth-century science and entertainment to pre-scuba undersea documentary films. The endpoint of her 2015 *The Mysterious Science of the Sea, 1775–1943* is the year of Cousteau's first underwater film shot with scuba, *Sunken Ships*. Shin Yamashiro includes undersea film in a long American tradition of sea writing in *American Sea Literature: Seascapes, Beach Narratives and Underwater Exploration* (2014). Writing about contemporary narrative films with underwater sequences, Adriano D'Aloia speaks of "enwaterment," as an aesthetic in films that "both physically and psychically engage . . . the spectator in a 'water-based relationship,'" whether or not their story concerns oceans or lakes.[37]

Ann Elias was the first to recover the fascination of the undersea for surrealism, which recognized the visual power of D'Aloia's "enwaterment," even in imagery made far from the sea. In *Coral Empire: Underwater Oceans, Colonial Tropics, Visual Modernity* (2019), Elias offers a compelling model for how to address the impact of submarine aesthetics on art and film by using the case of a single type of environment: mesmerizing coral reefs. *Coral Empire* opens paths as well in its attention to the first-world colonialism and racism that shape both the image capture and portrayal of underwater scenes. In a history inaugurated by the films of Williamson in the Bahamas, Elias contrasts the labor of people of color who participated in filmmaking, notably in diving and other water work, and their erasure in the finished products, which exhibit reefs as virgin territories for spectatorial delight and first-world tourism.

Since the late 2010s, scholars have started to recover the untold stories of the role played by both African diaspora and Indigenous divers of color across the history of maritime globalization, beginning in the early modern era, notably in salvage and pearl diving, as Rebekka von Mallinckrodt describes.[38] The great poet Derek Walcott evokes the work of Black divers in the Caribbean salvage industry in the haunting section "Rapture of the Deep" in "The Schooner *Flight*" (1979). There, the protagonist, Shabine, adds to his credentials as emblematic sea worker of color a stint diving with "a crazy Mick, name O'Shaugnessy, and a limey named Head."[39] As I have researched production on underwater film sets, I have found details showing colonial racial hierarchies abound, notably in warm-water locations preferred for submarine filmmaking, from Williamson's Bahamas studied by Elias, going forward into the scuba era. Thus, for example, to shift from the Caribbean to the Red Sea, while in Sudan filming *Under the Red Sea*, Hans Hass was attacked by a shark, although he did not include the episode in the film. Hass recounted that the attack took place while the Sudanese skipper, Mahmud, stereotypically portrayed in the film as jovial and carefree, was napping while Hans and Lotte Hass dived. When Mahmud awakened and realized that Hass has been bitten, he rowed over, and Hass "handed over the shark [that Hass was "pulling . . . after" him] to Mahmud. He hauled it up and battered its head in with a cudgel."[40] Such evidence affirms the importance of recovering the racialized distribution of labor in underwater filmmaking, as in tropical tourism industries more generally, although I do not open up this essential, challenging subject, attending exclusively to the technological capture of submarine materiality and its contribution to the realm of fantasy. Furthermore, underwater fantasies considered in this study have shaped the imagination of audiences beyond first-world contexts. Tellingly, Walcott repurposed nitrogen narcosis and Cousteau's term "rapture of the deep," along with cinematic vistas of coral, when he conceived of the strange emotions and seascape accompanying Shabine's delirious encounter with the bones of men and women who drowned during the Middle Passage.[41]

The propensity of the underwater environment to displace categories and concepts from land has been a recurring theme in the work of blue humanities scholars such as Mentz and Alaimo. Elizabeth DeLoughrey has recently used the concept "sea ontologies" to conceptualize this submarine defamiliarization of our land-based paradigms of aesthetics, philosophy, and history in her contribution to a recent influential section on "Oceanic Routes" in the journal *Comparative Literature* (2017). DeLoughrey illustrates "sea ontologies," analyzing the underwater art installations of Jason deCaires Taylor. Thus, she cites Taylor, noting how the art object's stability erodes underwater owing to the harsh environment: "underwater you are dealing with a completely different notion of time and it's confusing but fascinating; even for me when I return

after a month . . . the work looks completely different."[42] In 2019, the journal
English Language Notes published a special issue edited by Laura Winkiel, exploring both undersea and topside the "new perspectives, ontologies, and
transmaterial subjectivities" offered by oceanic frameworks.[43] Articles in this
issue, like the most recent work in the blue humanities, disrupt terrestrially
oriented concepts of scale and substance, as well as a human-centered notion
of the cosmos.

As Mentz and more recently Alaimo have emphasized, scholars need to
locate sea ontologies in the qualities of specific marine environments, if such
notions are more than metaphorical, anthropomorphic projections.[44] The
importance of situating both marine fantasy and marine reality in relation to
specific locations is a current trend in oceanic studies, as it is in environmental
documentary. Along with what might be called *placing* sea ontologies, another
challenge is to situate them in time. Thus, humans' destructive impact on
oceans in the age of the Anthropocene is at the forefront of critical, documentary, and creative interest in marine environments today. Such current awareness intersects with a long-standing history in Western aesthetics of imagining
the alluring, beguiling impact of the depths on the human senses, which Killian
Quigley and I have formulated as the senses of the submarine in the introduction to our recent edited collection *The Aesthetics of the Undersea* (2019).[45]

The human senses are very much at sea on the underwater film set as they
contend with altered perception and orientation, among the many formidable
aspects of submarine materiality. Throughout this book, the underwater environment is at once a remote, toxic realm and an opportunity for innovative,
enduring contributions to the repertoire of cinematic imagery and to fantasy.

1

Vision Immersed, 1840–1953

1.1. The Aquarium Is Not the Underwater Eye
(Immersive Vision, 1840–1890)

In May 1853, the first public aquarium opened at the London Zoo. The glass tanks in the zoo's Fish House may have been small, but they swarmed with life. There, visitors observed "58 species of fish and 200 invertebrates including 76 species of mollusc, 41 crustaceans, 27 coelenterates, 15 echinoderms, 14 annelids as well as a sprinkling from the small invertebrate groups."[1] Viewers were enthralled. Of course, people knew fish and shellfish as food, and they could peruse specimens classified on the pages of naturalist books, with every scale and fin crisply drawn. Middle- and upper-class vacationers had perhaps caught fish or collected shellfish, caught up in the century's growing enthusiasm for beachside leisure, even if they usually lived far from the sea. How different, it was, nonetheless, to observe water-dwelling creatures alive in their fluid element, *mobilis in mobili*, to cite Jules Verne's device for his imaginary submarine named after a magnificent cephalopod, the nautilus.

A year later, Philip Henry Gosse, the designer of the London Zoo's Fish House, published *The Aquarium: An Unveiling the Wonders of the Deep Sea*. In this book about aquarium keeping and specimen collecting, he evoked viewer pleasure in the aquarium's underwater spectacle, such as the lady marveling at the "pellucid" bodies and "smooth gliding movements" of prawns, leading her to compare them to "ghosts."[2] *The Aquarium* provided details to encourage amateurs to keep home aquariums, which proliferated in ensuing years, as did guides such as Gosse's, directed to crossover audiences, women, and also children. By the later nineteenth century, public aquariums became fixtures in major cities and resorts, and pop-up tanks drew crowds in transatlantic international exhibitions, most famously world's fairs.

The Paris World's Fair of 1867 had two grand saltwater and one freshwater installation among the attractions that drew between 11 million and 15 million viewers. The aquarium spectacles, like all of the exhibition, were the talk of the transatlantic world. Jules Verne and Hans Christian Andersen were among many writers and artists inspired by what one German contemporary called its "'magical castle on the ocean floor,' revealing 'secrets of the frigid deep.'" While aquariums had initially been developed to aid marine biologists, they "stand among the most successful medializations of the sea," from the Victorian era into the present day.[3] As Natascha Adamowsky concisely and lyrically observes, "the aquarium is a dream of the ocean given material form."[4] Curators starting with Gosse arranged aquariums with one eye on keeping animals alive and the other on attraction, drawing inspiration notably from the decorative arts. In containing real marine creatures arranged to tempt the imagination, the aquarium is an exhibition form at the threshold where biology meets fantasy, and in this way, it belongs to the same lineage as underwater cinema, which, as I have mentioned, also troubles the documentary/imaginative divide.

Yet another way aquariums resemble underwater film is that curators organized their novel spectacles drawing on aesthetics inspired by the submarine environment that predated modern access to the depths. Three aesthetics stand out. The framing of installations around prized specimens hearkened back to the cabinets of curiosities, which displayed rare, beautiful shells that had started to flow into Europe from expanding overseas trade.[5] Another inspiration was the baroque grotto, also dating to the early modern era when the arts were so powerfully shaped by Europeans' seaward ambitions. In baroque fantasies, classical water deities fused into ornate, rustic shelters, often ornamented with shells and rocks. The most magnificent of these installations bathed the environments in a humid atmosphere, thanks to early modern hydraulics.[6] Aquarium exhibitions also drew on the more intimate rococo style that was an inheritor of the baroque in the eighteenth century. Cultivating whimsy and irregularity, this aesthetic, too, took inspiration from the sea. It was "often described in terms of liquidity" and featured "shelly, florid, scrolling forms that seem to melt or flow seamlessly from one to the next."[7] All three of these aesthetics—the cabinet of curiosities, the baroque, and the rococo—shape, in miniature, for example, Gosse's installation for the Aesop prawn, as it appeared in a chromolithographic illustration for *The Aquarium* (fig. 1.1). Set off like a prized specimen (although common in coastal waters), the jewel-like prawn is nestled in a tiny grotto taking the form of an irregular arch decorated

FIG. 1.1. "The Aesop Prawn, &c." from Philip Henry Gosse, *The Aquarium: An Unveiling of the Wonders of the Deep Sea* (London: John Van Voorst, 1954), plate 6, p. 222. Photograph © The Royal Society. Gosse's installation frames this cold-water shrimp native to British waters, *Pandalus montagui*, as an exquisite specimen in a miniature baroque grotto.

with sea stars. On its head, what Gosse called the "slender filiform appendages" are, in his words, "continually thrown into the most graceful curves," and its delicate legs rest lightly on a paving of irregular barnacles, tinged in complementary pastel colors.[8]

In his preface, Gosse noted that he had heard about an "elaborate . . . scheme" of a French zoologist, who, in order to study marine life, "had provided himself with a water-tight dress, suitable spectacles, and a breathing tube; so that he might walk on the bottom in a considerable depth of water, and mark the habits of the various creatures pursuing their avocations."[9] Gosse expressed skepticism about whether such a dive was "really attempted," predicting "feeble results," even if it was, and reassured readers that "THE MARINE AQUARIUM . . . bids fair . . . to make us acquainted with the strange creatures of the sea, without diving to gaze on them."[10] Gosse thereby dismissed the investigations of the first scientific diver, Frenchman Henri Milne-Edwards, who published the observations he made when diving off the coast of Sicily in 1844.[11] Gosse's lack of interest was shared by other scientists as well by general transatlantic publics in the middle decades of the nineteenth century in a lapse so remarkable in hindsight that Adamowsky calls it "hydrophobia."[12] Despite such lack of recognition at the time, Milne-Edwards was pioneering a transformative method of submarine observation, with the diving helmet that provided the layer of air the human eye needs to focus, along with enabling breathing below the surface.

This layer of air was not available to free divers noted for their prowess since antiquity, such as sponge divers of the Mediterranean. European mariners were famously averse to swimming. When Europeans landed in the Americas, they were astounded by the ease in the water of Indigenous peoples. According to Robert Marx, "[t]he Spanish historian, Oviedo," writing in 1535, was one among many who "marveled at their abilities, saying that they could descend to depths of nearly a hundred feet, [and] remain submerged as long as 15 minutes."[13] Similar capacity characterizes traditional free divers on the other side of the globe, most famously Ama, the Japanese pearl fisherwomen. Recently, scientists have coined the term "sea nomads," to describe the Bajau peoples of the Southeast Asian archipelagoes, who have physiological adaptations enabling them to spend remarkable periods of time (up to thirteen minutes) underwater with a single breath.[14] Perhaps Oviedo's observations half a millennium ago about the breath-holding capacities of Indigenous divers in the Americas were not a travel writer's exoticizing exaggerations. In any case, the difficulty in focusing the eyes underwater, along with lack of air, limits what the free diver can observe. The best technology available for providing the eye with enough air to focus in the premodern period was translucent shells used as coverings.

Through the helmet's layer of glass, nineteenth-century divers looked at a planetary environment that had never before been observed, let alone described. A version of the diving bell was improved for observation as well, when in 1865, French engineer Ernest Bazin devised a diving chamber "of high grade steel" with "two viewing ports and an external electric lamp for underwater illumination" to "search for a treasure sunk in Vigo Bay, Spain."[15] In 1875, an Italian engineer enhanced the diving chamber's safety by supplying compressed air that would refresh the atmosphere. Yet these technologies were, until the 1880s, almost exclusively used for industry. "No one questioned the incomparable value of diving or the wholly new opportunities and fields of research it opened," yet "one was unlikely to encounter interest in underwater landscapes or lyrical effusions"—in sum, descriptions of the environment based on observations while immersed.[16]

What did transatlantic publics know about submarine conditions in the era between the inception of modern industry diving and the invention of the photosphere? As a prelude to exploring how filmmakers shaped the submarine realm for publics, I review in this section what details were disseminated during this era, using the scant secondary literature on the subject, which directed me to a few primary examples as well. The picture emerging from this survey is that the material conditions of the underwater environment were almost entirely unknown to general audiences at the time. Such lack of knowledge is important context because it underlines how unhampered filmmakers were by viewer expectations once they had technology to film below. This lack of expectations was at once freeing and also provided a challenge. While creators could shape imagery according to their visions, at the same time, this imagery had to convince audiences that it was indeed the undersea environment, a challenge all the greater because the environment was inaccessible to them.[17] Leisure diving would not develop until the advent of scuba in the post–World War II era.

Writing about diving in Sicilian waters in 1844, Henri Milne-Edwards commented, "I saw perfectly everything that surrounded me."[18] From the murky waters most often plied by industry divers of the northern Atlantic, the reports that came to the general public were of underwater gloom. In *The History of Underwater Exploration*, Robert Marx quotes in full the anecdote from an anonymous amateur who dove in the Thames in 1859. This viewer compared the underwater ambience to "the London atmosphere in a November fog," with "living forms floating about here and there," although he could not "say what they were."[19] A year later, Charles Dickens, from the fog-filled bank at night, rather than the murky river, wrote "methought I felt much as a diver might, at the bottom of the sea."[20] The underwater gloom was more enjoyable for Robert Louis Stevenson, in his brief firsthand account of a dive, which he made in 1868, published in 1888. The black sheep of a noted British family of lighthouse and civil engineers, Stevenson described how as a young man he visited a construction site on the Bay of Wick and persuaded a "certain handsome scamp of a diver," Bob Bain, to let him experience helmet diving.[21] Stevenson describes how Bain led him down a ladder into the murky depths and how strange it was to discover an atmosphere where he could fly; the pressure on his Eustachian tubes; and also "a fine, dizzy, muddle-headed joy in my surroundings." He experienced the dimness below as "a green gloaming, somewhat opaque but very restful and delicious," although Stevenson also acknowledged himself "relieved" when he was summoned up by Bain from this bizarre, if fascinating, experience.

By the time Stevenson published this essay in 1888, knowledge of underwater stress on the human body was widespread. One impetus to such realization was the diagnosis in the early 1870s of a new disease that crippled and even killed submarine workers in what were called "caissons": chambers for working on feats of submarine engineering, such as building the Eads Bridge in Saint Louis and the Brooklyn Bridge in New York in 1870. Scottish sea fiction and adventure writer R. M. Ballantyne was the first novelist to portray the workaday dangers of helmet diving in *Under the Waves or Diving in Deep Waters* (1876). The website Classic Dive Books identifies *Under the Waves* as "the first adventure book written about diving for young readers," and it may be the first adventure novel about diving based on direct observation for any readership.[22] In the preface, Ballantyne explains that he consulted with professionals (whom Verne had only named in *Twenty Thousand Leagues*, published six years before). Ballantyne thanked "Messrs. Denayrouze, of Paris for permitting me to go under water in one of their diving-dresses."[23] He also wrote, "I have to acknowledge myself indebted to the well-known submarine engineers Messrs. Siebe and Gorman, and Messrs. Heinke and Davis, of London, for much valuable information."[24]

In this novel, gloom is the ambient atmosphere below. The glow starts out as ominous. When the character Rooney makes his first dive, struggling to orient himself, he is surprised when a dead cat, which "had not been immersed long enough to rise to the surface," comes into view. It "floated past with the tide, and its sightless eyeballs and ghastly row of teeth glared and glistened on him, as it surged against his front-glass."[25] The gloom becomes more routine but no less sinister when the hero, Edgar, descends to salvage the *Seagull*, which had shipwrecked while bound for the British colony at the Cape of

Good Hope and had sunk in ten fathoms of water (about sixty feet), off the coast of England. In such depths, Edgar has technical assistance and "looked with much interest at his little lamp." "The water being dense and very dark its light did not penetrate far, but close to the bull's-eye it was sufficiently strong to enable our hero to see what he was about."[26]

In contrast to his vague portrayal of the atmosphere, Ballantyne gives a detailed depiction about the workaday dangers confronting the diver. These include obscurity, difficulties using compressed air, and the pressure of "the ever-increasing density" of the atmosphere on his body.[27] Ballantyne portrays such difficulties from the moment Rooney takes his first tentative steps into the depths and must figure out how to maintain neutral buoyancy—so as neither to rise nor to sink in this heavier atmosphere. As he starts to rise too fast because of the air filling his suit, he feels "a singing in his head and a disagreeable pressure on the ears."[28] Without instruction but using his wits, he quickly figures out how to solve this problem by opening a valve to release excess air.

Ballantyne was a writer of popular sea adventure fiction, and with *Under the Waves*, he does, indeed, take topside sea fiction below. In doing so, he follows sea adventure fiction's practice of using technical details of a profession to create problems and solutions involving readers in the plot, which I have described in *The Novel and the Sea*. As is the case with conventional sea fiction, Ballantyne only describes aspects of the submarine environment as they are part of work, whether it is finding one's way in the darkness or moving through the heavy atmosphere.

The description of the environment as it figures in work also characterizes industry descriptions of diving, if the anecdotes in Sir Robert Davis's *Deep Diving and Submarine Operations: A Manual for Deep Sea Divers and Compressed Air Workers* are representative. This book is an authoritative overview of professional diving by a man who worked for Siebe Gorman from the age of eleven and rose through the ranks during the first half of the twentieth century to become its director.[29] In part 2 of *Deep Diving and Submarine Operations*, Davis includes a survey of some of the most famous—sometimes heroic, sometimes fatal, always arduous—industry dives across history, drawing on accounts from the time, in a number of cases based on firsthand observations. The dives surveyed range from early modern salvage efforts immediately after the loss of the HMS *Mary Rose* in July 1545, to the 1920s salvage of German ships scuttled at Scapa Flow shortly after World War I, undertaken by Cox & Danks Ltd. company. In the twentieth century, Davis also includes Beebe's naturalist dives. Davis's classic book was reprinted a number of times, and by the 1955 edition, incorporated stories of salvage hardhat diving into the post–World War II era.[30]

If industry accounts contained so little information about the submarine environment for its own sake, however, it was for a very different reason than the "hydrophobia" that Adamowsky attributes to Victorian scientists—because divers did not need to recapitulate its features in work written to the craft. Satire often reveals the truth of convention, and thus a shorthand for industry conventions at the time Siebe innovated his helmet was a parody in *Punch* in 1843. *Punch* targeted the salvage of *Le Télémaque*, a ship lost on the Seine, that was supposed to contain fabulous wealth. The salvage was written up by *Le Moniteur* and summarized in the *Mechanics Magazine*. The *Punch* satire called out the "classic terseness of the style of [leading helmet diving engineer and marine salvagers] Deane and Edwards," and concluded with the "startling declaration that there is positively no treasure on board the *Télémaque*," after "divings and draggings, and drenching and duckings . . . in the vain hope of fishing up something from the wreck of the ill-fated vessel."[31]

A serious example showing how the submarine environment appears only as it impacts labor was an account of salvaging the HMS *Eurydice*, which had sunk off the Isle of Wight on March 24, 1878, with three hundred lives lost. When the wreck was raised that summer, principal challenges were "that the men could only go below at slack tide and in good weather, that the currents were very strong, and that the wreck was embedded in 9 or 10 feet of mud."[32] Portrayals of divers struggling with such currents, or the quality of the mud are absent. Instead, Davis explained and also offered diagrams both of how the wreck lay on the sea floor, and "the process of pinning down" at low tide: "the ropes were hauled taut and made fast to the lifting vessels, so that when the tide rose, the lighters and their burden rose too." A painting celebrating the salvage was given to the London Science Museum by Siebe Gorman—namely, Charles Roberts Ricketts's *HMS "Eurydice" under Salvage* (1882) (fig. 1.2). Ricketts's image has the panoramic sweep of Riou's promenade through the forest of Crespo for Verne's *Twenty Thousand Leagues under the Seas*. Thus, Ricketts gives activity in the depths the same amplitude as topside action, fusing conventions of maritime painting with information characterizing a diagram of undersea work, rather than offering a realistic depiction of the scene. Such grand vistas disappeared in imagery based on firsthand observation below, even observation as limited as Wyllie's in preparing *Davy Jones's Locker*.

FIG. 1.2. Charles Robert Ricketts, *HMS "Eurydice" under Salvage* (1882), oil on canvas. Photograph © Science Museum/Science & Society Picture Library. Ricketts creates a magnificent panorama that joins work conducted at the surface with work underwater, applying the sweep and scale of landscape as viewed through air to optical conditions in the depths.

Amidst the era's pervasive hydrophobia, Adamowsky notes the extraordinary prescience in the 1860s of Austrian naturalist Baron Eugen von Ransonnet-Villez, who became captivated by the beauty of tropical corals. Ransonnet published two travel books with the first extended descriptions, and also the first visual images, based on prolonged, firsthand observation in the Western tradition. For *Travels from Cairo to Tor to the Coral Reefs* (*Reise von Kairo nach Tor zu den Korallenbänken*) (1863), Ransonnet was free diving. Even with his limited time below, he noted submarine luminosity and the behavior of color: "How peculiar things appear under water! Though one cannot exactly distinguish the contours in the deep, yet everything gleams in beautiful and strange illumination! Brown, violet, orange, in yellow and blue light, everything glows towards the diver."[33] In the years after publishing this account, Ransonnet designed a custom diving bell with a window so that he could sketch below. He used this diving bell for his trip to Ceylon (present-day Sri Lanka) and included both verbal descriptions and engravings in *Sketches of the Inhabitants, Animal Life and Vegetation in the Lowlands and High Mountains of Ceylon* (title of the 1867 English edition, plates and captions only; expanded narrative published in German in 1868). There, for example, Ransonnet observed again the details of altered visual perception below. "Strange seemed the light effects down there in the sea so I paid special attention to it. Bluegreen is the basic tint of the underwater landscape and especially of all bright objects, whereas dark, e.g. blackish rocks and corals, and far away shadows, seem to be wrapped in a monotone maroon, which is in complementary relation to the colour of the water."[34] Despite such novel observations, these works "remarkably . . . did not command much attention at the time," with a "small print run."[35]

From the scant information in secondary literature, it seems that both scientific and public interest in submarine reality started to crystallize in the 1880s–1890s. Within this time frame, historian of scientific diving and underwater photography Hermann Heberlein names several noteworthy scientists who turned their attention to underwater optics.[36] The most famous was biologist and artist Ernst Haeckel, who knew of Ransonnet's depictions. In *Nature*, Haeckel published an article where he lamented his lack of access to such a diving bell. Nonetheless, he commented that by training his eyes to remain open, he could observe "the mystic green light in which the submarine world was bathed, so different from the rosy light of the upper air. The forms and movements of the swarms of animals peopling the coral banks were doubly curious and interesting thus seen."[37]

Marine biologist Hermann Fol, a student of Haeckel's, realized that such conditions were worth attention in their own right. In an article from 1890, based on his experience diving in the Mediterranean, Fol described underwater optics and tied them to two practical purposes—submarine navigation and underwater photography. While the shallow depth of field thwarted sight for navigating undersea vessels, Fol was optimistic about underwater photography. He noted the loss of red light and surmised that the blue rays that last the longest are, in Fol's estimation, "the rays that act with the greatest energy on the photographic plate."[38] Fol also noted the altered submarine color spectrum, the effect on visibility of different angles of the sun, and varying turbidity of water in different zones. (His comments about poor underwater visibility also raised for him the question about whether fish were nearsighted: "what use would distance vision be, because in any case, they would only be able to see several meters?")

In 1890, when Fol published his observations, experimentation for developing reliable processes for underwater photography was under way. French marine biologist Louis Boutan is credited by photographic historians with the first clear, reliable underwater photography. In 1900, Boutan outlined his method thoroughly in the book *La photographie sous-marine et les progrès de la photographie*. Boutan's precursors included William Thompson, who took an exposure in Weymouth Bay in February 1856, as well as German submarine inventor Wilhelm Bauer, and the previously mentioned Frenchman Bazin who upgraded the diving chamber.[39]

Slides of Boutan's photos were shown at the great Paris World's Fair of 1900. By this time, curiosity, if not knowledge, about submarine conditions was growing in the general public. Young adult literature about diving began to appear in English in 1887, a decade after Ballantyne's *Under the Waves*. Such books are listed on the website Classic Dive Books, dating to Harry Collingwood's 1887 *The Log of the Flying Fish: A Story of Aerial and Submarine Peril and Adventure*.[40] Publics were fascinated by the real-life success of Alexander Lambert, "who had recovered the vast majority of gold bullion from the 1885 wreck of the *Alphonso XII* in the Canaries."[41] A particularly successful melodrama on the London stage in 1897 was Cecil Raleigh and Henry Hamilton's *The White Heather*, culminating in an underwater fight scene represented in advertisements for the production, which was popular enough to cross the Atlantic to Broadway. H. G. Wells noted the change in color of the sea in portraying the dive of a submersible into the abyssal depths inhabited by aliens in his short story "In the Abyss" (1896), which was one inspiration for James Cameron's *The Abyss* (1989). As the protagonist, Elstead, plunges downward, he "saw the water all around him greeny-blue, with an attenuated light filtering down from above, and a shoal of little floating things went rushing up past him. . . . [I]t grew darker and darker, until the water above was as dark as the midnight sky." Further, "little transparent things in the water developed a faint glint of luminosity," as they "shot past him," suggesting bioluminescence.[42]

If this time frame correctly identifies the intensifying public curiosity about submarine reality, it coincides with the invention of underwater photography. Did general interest in the environment lead inventors to take photography below? Did public curiosity grow as submarine photography revealed the unique qualities of submarine materiality? Or, as is often the case, were public attention and technological affordance mutually enhancing? Certainly, a knowledge of aquatic physical properties was required for an underwater camera to succeed. Boutan gives evidence of how closely he was studying underwater atmosphere, including its movement and its light in an 1894 article on his process.[43]

As Heberlein observes, Boutan was not the only inventor in the 1890s, which launched an intense period of innovation in underwater photography. "At about the same period as Boutan—1891—Paul Regnard, a professor at the Sorbonne, was also engaged in underwater photography. Just before the turn of the century, in 1898 and 1899, J. E. Romborsts . . . and C. L. Bristol . . . photographed under water. Etienne Peau was also an excellent underwater photographer." The most creative name for the underwater apparatus was coined by de Casparis: the "Thalassiuscope," although it is not clear whether it "was ever employed."[44] In the first decade of the twentieth century, British naturalist Francis Ward built a fish pond with an observation chamber that he could use for study and photography, as he detailed in *Marvels of Fish Life as Revealed by the Camera* (1911). In 1912, American J. E. Williamson invented the photosphere, which he modeled on a technology for salvage and ship repair designed by his father. This technology was "a deep-sea tube, made of a series of concentric, interlocking iron rings that stretched like an accordion" and could be suspended from a ship, allowing communication as well providing an air supply. "When attached to a diving apparatus, the tube could assist in underwater repair and salvage work."[45]

Because, as I have suggested in this section, there was little knowledge of underwater conditions circulating among the transatlantic public before the Williamson photosphere, they did not come to underwater film with a strong imaginative template. There are few other cases—the extreme environments of the polar oceans and the highest mountains on earth are others that come to mind—where we can see so clearly the interplay of technologies and the imagination at work in shaping a physical environment into imagery that is both entertaining and that persuades publics that the imagery is capturing the environment's reality. In the case of the submarine realm, factors in this interplay include the type of access afforded by technologies, the environment's strange and beguiling qualities, these qualities' match with preexisting aesthetics as well as dominant aesthetics of the time, and the expectations of publics whose attention filmmakers seek to capture. A particularly important factor informing the creation of imagery is how the submarine realm defied expectations publics brought from the surface. These expectations included their basic habits of seeing through air and, in the case of Western publics, their expectations about representations of landscape based on conventions dating back to the Renaissance.

1.2. Seeing through Water: Aquatic Perspective
as Symbolic Form

In the following section, I explain how underwater optics diverge from viewer expectations, showing the problems they pose to Western pictorial conventions for depicting landscape. My artifacts for this explanation are imagery by the first two artists to depict submarine environments based on extended time below, using industrial diving technology. One of these artists is the previously mentioned Baron Ransonnet-Villez, who drew from a diving bell, including images published in his *Sketches of the Inhabitants, Animal Life and Vegetation in the Lowlands and High Mountains of Ceylon* (1867 English, 1868 German). From Ransonnet, I turn to the paintings of Walter Howlison "Zarh" Pritchard, who made an entire career from observations in the depths. Pritchard was born in Madras in 1866 to Anglo-Irish parents. He was educated in Scotland, where, a century before modern wet suits and dry suits, he started diving in the frigid Firth of Forth, using goggles he improvised from cowhorn and leather.[46] Pritchard had an adventurous and unconventional life, from his childhood in Scotland to his sojourn in Tahiti in 1904, when he first essayed helmet diving. At the height of his career, Pritchard was based in Bishop, California, and traveled to dive, record, and sell his pictures around the world.

While forgotten today, Pritchard's works were popular in the first quarter of the twentieth century, a transitional era when publics were becoming interested in the submarine realm, accompanying its first representations in photography and film. Pritchard's scenes interested scientists and were acquired by institutions with both professional and popular missions, such as the Scripps Institution of Oceanography in La Jolla, the American Museum of Natural History in New York, and the Cleveland Museum of Natural History. Prince Albert I of Monaco, who undertook oceanographic voyages across the globe, purchased eleven paintings by Pritchard for the Musée Océanographique in Monaco after seeing a Pritchard exhibition in 1921 at the prestigious Galerie Georges Petit in Paris, where Pritchard would have a second show in 1925. Pritchard's paintings were also enjoyed by fine arts connoisseurs. Lists of collectors include the royalty and the wealthy in Europe, the United States, and Japan, including theatrical celebrities like Sarah Bernhardt and Ellen Terry, and his paintings were acquired by the Musée du Luxembourg in Paris, the Boston Fine Arts Museum, and the Brooklyn Museum, among others.

When interviewed, Pritchard explained that he was able to sketch if not paint undersea with "a special method," using "calfskin leather" for a canvas.[47] "The canvas having been previously covered with linseed oil, the thick oil colors adhered to it perfectly, having the property of not dissolving in contact with the water."[48] Pritchard gave an evocative description of some challenges specific to the environment, including a "tiny mistake [when] . . . a fish has come too near, and the whole sketch is spoiled. A submarine painter can make no alterations."[49] How many of Pritchard's paintings were in fact done underwater is unclear—many were probably at least finished on land, if not drafted entirely from preliminary underwater sketches.

There are two reasons that I use hand-drawn imagery by Ransonnet and Pritchard for laying out the difficulties of adapting underwater perception, and above all optics, to prevailing Western landscape conventions in a book about underwater film. If Louis Malle remarked of filmmaking after the invention of scuba, "we had to invent the rules—there were no references; it was too new," his point is even more the case for the first two painters to paint submarine scenery based on prolonged immersion.[50] Their art shows the pressures of adapting viewer expectations formed topside to submarine conditions, and the creative possibilities unleashed by attention to their specificity. I also use these two artists because they worked without the mediation of the camera's lens, which has been most commonly designed to further traditional Western conventions at odds with definitional features of underwater perception. From the perspective of underwater camera and dive innovator Hans Hass, who figures centrally in part 2, "Ransonnet captured the defining quality of being underwater in a way that modern photography did not afford."[51]

Throughout this section, I describe how Ransonnet and above all Pritchard organized the underwater environment as perceived by the immersed eye into pictorial representation. In doing so, they created a new symbolic form that I term "aquatic perspective."[52] With the notion of "symbolic form," I echo art historian Erwin Panofsky's *Perspective as Symbolic Form*, a foundational essay both describing and historicizing conventions for organizing three-dimensional space on a two-dimensional plane. Panofsky influentially argued that the visual organization of space in an image is not a transparent window onto reality but an expression of a historical worldview. Panofsky's analysis attends to Renaissance linear perspective, shaped variously, aesthetically and epistemologically, by a humanist worldview that places the observer in a position of mastery, where seeing is associated with the power to control.

Ransonnet and Pritchard, in contrast, organize the strange (to the terrestrial eye) qualities of underwater space in such a way that the equation of seeing with mastering starts to dissolve. Their achievements tacitly acknowledge that industrial technology lets them dive, yet at the same time, their vistas reveal a realm of disorienting spatial relations, muted, arresting colors, and the unfamiliar life of an atmosphere where humans are not at home, even as it covers large areas of our planet. Such displacement of humans, *not from knowledge but from mastery*, owing to the material qualities of earthly environments rather than any divine presence, is a defining feature of the new symbolic form coalescing in Ransonnet and Prichard's undersea imagery.

My use of the term "aquatic perspective" also echoes William Beebe's *Beneath Tropic Seas: A Record of Diving among the Coral Reefs of Haiti* (1928), one if not the first book entirely devoted to underwater dive naturalism, describing marine flora and fauna, as well as underwater optics, movement, and color.[53] For Beebe, "one artist, Zarh Pritchard, has brought to canvas evanescence of hue, tenuousness of tint eminently satisfying to the memory of the stroller among coral reefs. This is probably because he paints under water, seated among his subjects."[54] Beebe, indeed, used a painting by Pritchard, *No Man's Land Five Fathoms Down*, painted in the lagoon of Maraa, Tahiti, as the frontispiece to the first edition of *Beneath Tropic Seas*, a book otherwise illustrated with photographs, including enlargements of photos taken with the Williamson photosphere. In praising Pritchard's work, Beebe distinguished it for exhibiting "aquatic perspective" that "no aquarium tank can ever show"; "[n]o glass-bottomed boat ever conveys the mystery and beauty of this under-world of color" that Pritchard expressed so well.[55] With "aquatic perspective," Beebe was specifically describing "the pastel film" of Pritchard's surfaces.[56] I will extend his term to describe the pictorial organization of underwater scenes expressing their perception by the immersed human eye, in tandem with the immersed body.

Both Ransonnet and Pritchard have fallen outside the mainstream of visual studies, and underwater landscape was certainly not a common subject for nineteenth- and twentieth-century painters. Art historian Charles Eldredge is one of the few to survey its history, yet he downplays the interest of submarine materiality. As Eldredge writes, "[t]he scientist's empirical approach was, however, of scant concern to many modern artists who were drawn to submarine motifs not because of new knowledge, but because of old mysteries."[57] Accordingly, Eldredge does not recognize that to understand imagery, it is critical to take account of the state of the technology for underwater access, although he does comment on how Charles Demuth was "moved to new aesthetic experiments by his underwater experience at the Battery [aquarium]."[58]

Aquariums revealed the movement of living sea creatures, but they exhibited little else about immersive perception. Once in water, humans enter an atmosphere eight hundred times more dense than air. As a result of this density, when we look through water, we perceive a pervasive haze, because the atmosphere of water slows down light, absorbs it, and scatters it. Additionally disrupting clear vision underwater are "minute particles suspended in water," and also, at shallow depths, the way sunlight is refracted or diffused by the texture of the ocean's surface.[59] The furthest the eye can see in the clearest of water, distilled water, is theoretically 80 meters (262.47 feet). Martin Edge gives some striking figures on this reduced visibility in his practical guide, *The Underwater Photographer.* "Such is this density of water," he writes, "that people compare a picture taken in 1 m of water with a picture on land taken at 800 m away."[60] Because visibility is so radically diminished, objects fade away into the haze as they recede, much like our depth perception amidst fog in the atmosphere of air.

The density of water also explains the different color spectrum underwater. This atmosphere absorbs light quickly, making colors with weaker, faster light waves imperceptible as the diver descends away from the sun. Although the exact depth of color loss depends on a number of local conditions (how turbid the water is, the strength of the light above the water, and the turbulence of the surface), a chart from NOAA shows the typical penetration of light at different depths in open ocean and coastal waters, perceived without assistance from artificial lighting (fig. 1.3). Beebe was captivated by how underwater color defied his expectations. "Striving to fix and identify remembered hues of a coral grove, I lose faith in my memory when, in my color book, I find them listed as Russian blue or onion-skin pink."[61] He wrote, "We need a whole new vocabulary, new adjectives, adequately to describe the . . . colors of under sea."[62]

The pervasive haze and the altered color spectrum are just two disorienting aspects of seeing through water. Our depth perception is further confused because phenomena are magnified by about 33 percent, and stereoscopic vision does not work well. Helen E. Ross explains that underwater, as in mist seen through air, "the low contrast of the images makes the disparity difficult to detect, and stereoscopic acuity is much poorer than would be predicted on the grounds of retinal image sizes alone." This is particularly true close up, when the brain experiences a conflict between the tactile knowledge of distance and underwater magnification. But it is also true in distances beyond our reach,

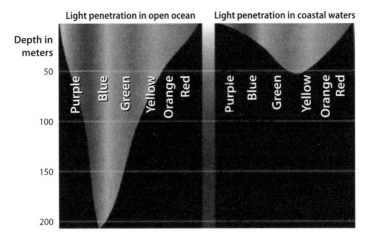

FIG. 1.3. "Light Penetration in Open Ocean / Light Penetration in Coastal Waters." Kyle Carothers, NOAA-OE. The effect of depth on human perception of color underwater, when unaided by artificial light.

because, as Ross points out, stereoscopic perception depends on "the disparity in the images received by the two eyes [which] is a fairly powerful cue to the relative distance of objects, and contributes a great deal to the sensation of three-dimensionality."[63] Ransonnet noted how underwater "one's normal sense of distance and size is completely lost": "eagerly I stretched out my hand towards a coral, but could not touch it, just like a child who tries to grasp things beyond its reach, because in the water everything seems to be so deceptively near and, at the same time, smaller." Ransonnet continued this observation with the memorable sentence: "You soon realize that in the depths of the ocean you need not only to learn to move, but how to see and hear as well."[64] In describing the feeling of being at depth, Pritchard observed that "[r]ocks and cliffs in the dim light assume an appearance of inconceivable size."[65] Pritchard also noted how hard it was to orient oneself: even touching bottom underwater was "as if one were temporarily resting . . . on a rapidly dissolving fragment of some far planet."[66]

"How can life be organized in a world without horizon?" asked Philippe Diolé, a member of the French navy team led by Cousteau that pioneered scuba in the 1940s, striving to convey a sense of our visual disorientation beneath the water's surface.[67] The same question might be transposed to the problem of rendering underwater in two dimensions as well. Seeing through water is not only at odds with seeing through air but also with Western conventions for depicting scenery since the Renaissance. Reduced depth of field, haze, and the distortions of one's "normal sense of distance" (Ransonnet) notably present a challenge to linear perspective.

The foundation of linear perspective is the organization of a "ground plane [that] . . . clearly permits us to read not only the sizes, but also the distances of the individual bodies arrayed on it."[68] Geometric forms are placed on this ground plane, and they recede proportionally. An architectural element like a pillar or checkerboard tiling, "becomes an index for spatial values, and indeed as much for those of the individual bodies as for those of the intervals."[69] The depiction organizes the scene along a visual pyramid, whose apex is a single vanishing point on the horizon where orthogonal lines meet. Underwater, the poor visibility thwarts linear perspective's recession to a vanishing point. While spaces in rooms or cathedrals might be compressed enough to dovetail with viewing a reef or other underwater scene, the closest terrestrial equivalent outdoors to underwater scenes would be viewing in a dense fog. Along with the mathematical organization of the two-dimensional picture plane, linear perspective assumes a monocular viewer, whose fixity is another challenge for underwater depiction using this convention, along with its reliance on straight lines. The tendency of the viewer underwater is to move, contending with surge and sway.

In an environment so at odds with linear perspective, Ransonnet and Pritchard experimented with how to convey qualities of submarine conditions without completely disorienting the viewers of their novel scenes. In introducing his imagery in *Travels from Cairo to Tor to the Coral Reefs* (1863), Ransonnet noted that "the illustrations presented here show the seafloor like on *the inside of a fish tank, in the same perspective arrangement as it is used in landscapes*. Only coral species are shown in one picture that are really found there, their position and their succession in different depths and on different seafloors were taken into account" (fig. 1.4). He thus acknowledges implicitly reigning public expectations both that the undersea resemble an aquarium and that it can be organized with conventions for depicting landscape honed in the atmosphere of air. At the same time, Ransonnet also remarks that the organization of the aquarium thwarts his naturalist goal: "unfortunately many genera had to be omitted which were mentioned in the article."[70]

Even working with such constraints, Ransonnet captures submarine magnification and the underwater haze, notably in images initially sketched from the diving bell. Take, for example, one of the images in the collection of the Oceanographic Museum of Monaco analyzed in detail by Jovanovic-Kruspel,

FIG. 1.4. Eugen von Ransonnet-Villez, "Coral Group in the Harbor of Tor." Photograph ©
Natural History Museum, Vienna. The image is used in Ransonnet's *Reise von Kairo nach Tor
zu den Korallenbänken des rothen Meeres* (Vienna: Carl Ueberreuter, 1863), fig. 1, p. 35, where it
is described as follows: "An outstanding group of corals, like an oasis out of the muddy sand,
in a place that is protected from the waves. During low tide, the seafloor is covered by about
six feet of water." The forms and colors of the scene convey aquatic perspective, yet Ransonnet
also frames and flattens the scene in keeping with the conventions of the aquarium.

Pisani, and Hantschk (fig. 1.5). A comparison of this image with the Aesop
prawn installation in Gosse's chromolithograph brings out the difference between the views offered by the aquarium and vision immersed. Ransonnet has
thickened the atmosphere, conveying a sense of its materiality. Compared to
the rounded quality of the coral heads in Ransonnet's images, Gosse's prawn
poses stiffly on the bottom of the scene, cut off from us by the wall of aquarium
glass, as are Gosse's barnacles. Ransonnet's corals, in contrast, with their beau-
tifully rendered, fine edges, enhance the scene's outreach to the viewer, and
their looming suggests underwater magnification.

    Although Ransonnet's images from the diving bell are discolored with age,
they still reveal how Ransonnet experimented with color, as well as line, to
invent conventions that will express seeing through water. Notably, Ransonnet
recognized that he could use another long-standing Western convention for

FIG. 1.5. Eugen von Ransonnet-Villez, "Two Groups of Madrepores off the Coast of the Island of Ceylon [present-day Sri Lanka] near Point de Galle Drawn from a Diving Bell, February 3, 1865" (Deux groupes de Madrépores sur la côte de l'Ile de Ceylan près de Point de Galle dessinés dans une clôche à plongeur. 3 février, 1865), oil on paper. Photograph © Institut océanographique, Fondation Albert 1er, Prince de Monaco. Corals loom in three dimensions toward the viewers in this exquisite scene drawn from a diving bell, conveying the enveloping sense of the liquid atmosphere.

creating a sense of three-dimensional space in two dimensions to organize the picture plane, amidst haze. "Aerial perspective" was the term employed by Leonardo da Vinci when he considered how buildings and mountains appear not only smaller but also blurry when seen in the distance. Furthermore, such remote features take on the color of the atmosphere. He thus advised artists to "make the building which is nearest . . . of its natural color, but make the more distant ones less defined and bluer," "when the sun is in the East."[71] Helen Ross notes that "[a]erial perspective is a much more important cue under water than on land—partly because the loss of contrast is more noticeable, and partly because other distance cues are reduced or absent."[72]

In Ransonnet's images from the diving bell, he translates aerial perspective under the waves. Changes in clarity and color render depth, with the modification that the viewer's depth of field is drastically limited, compared to air. Thus, the extraordinarily delicate edges of the front of the coral heads in the foreground of the picture start to blur as they recede into the background. Further, the coral head in the foreground of the picture has a greenish tint, differentiated from the brownish coral in the middle ground, which shades into the background. Its subtle coloring conforms to Ransonnet's previously cited description of underwater color in his *Sketches*: "Bluegreen is the basic tint of the underwater landscape and especially of all bright objects, whereas dark, e.g. blackish rocks and corals, and far away shadows, seem to be wrapped in a monotone maroon, which is in complementary relation to the colour of the water."[73]

Ransonnet employed both blurring and coloring to give a sense of depth underwater in his images included in the *Sketches*, as well as in *Travels from Cairo to Tor to the Coral Reefs*. This use is most evident in his images from the diving bell. In his published imagery, in contrast, aquarium conventions are more prominent. Thus, specimens are laid out with clarity in the foreground of the image, whether on the sand or on a reef. At the same time, aquarium conventions cannot resist the compelling revelations of immersive perception. Notably in the example from *Sketches*, the reefs in the back of the image take on the coloring of the water, and the hazy schools of fish add a sense of movement and mystery.

Stefanie Jovanovic-Kruspel, Valérie Pisani, and Andreas Hantschk appreciate "the almost photo-realistic and scientific representation" of Ransonnet's imagery.[74] Their assessment recognizes Ransonnet's documentary aspirations. After visiting Ceylon, Ransonnet "began to experiment with colour photography," which failed, however, because the collodion plates used in photography at the time were not sufficiently sensitive to light to capture imagery underwater.[75] At the same time, Ransonnet's attention to the specificity of submarine conditions creates a beguiling mood. In Adamowsky's characterization, Ransonnet's images stand out "for the singular atmosphere they radiate"; Ransonnet "conceived an aesthetics of submarine indeterminacy that represented the water's material resistance to penetration by the observing scientific gaze."[76]

In his preface to *Travels from Cairo to Tor to the Coral Reefs*, Ransonnet wrote, "*The submarine views drawn by me are only the first, weak attempt in this unprocessed field. The Claude Lorrain and the Rosa Bonheur of the underwater landscapes might not even have been born yet.*"[77] As Jovanovic-Kruspel, Pisani, and Hantschk comment, "Whereas Claude Lorrain was renowned for his romantic perception of his landscapes Rosa Bonheur the famous 19th century animal painter was praised for her highly realistic observations."[78] Few understood Ransonnet's excitement about the possibilities for submarine painting in his time—whether documentary or Romantic—and this environment would long remain what Ransonnet called "an unprocessed field" for painting.[79] Wyllie's *Davy Jones's Locker*, mentioned in the introduction, is an exceptional glimpse of the depths from a helmet improvised from a cookie tin. In contrast, Zarh Pritchard used a diving helmet to gather information for his visions of the depths, and hence Pritchard is the first painter to paint in *pleine mer*.

Perhaps in part because of his extensive time diving, Pritchard's aquatic perspective frees itself from aquarium constraints. Pritchard is fascinated by the vertical dimension of underwater space, and he renders reefs falling off into unfathomable depths rather than anchoring images with a ground plane, recalling an aquarium. Furthermore, Pritchard leaves the surface of the water out of his imagery, in contrast to the Ransonnet, which is another residue of aquarium orientation. Instead, Pritchard immerses viewers in an "all-enveloping atmosphere," where " not one object looks wet or glitters."[80] The softness of this atmosphere is an abiding preoccupation of Pritchard's, in contrast to aquarium viewing, where the glass containing the water can gleam and reflect.[81]

The softness of the underwater realm infused the muted quality of Pritchard's color palette. J. Malcolm Shick credits Pritchard with recognizing "the altered submarine spectrum" (without artificial lighting), and this altered spectrum is an important aspect of Pritchard's aquatic perspective.[82] In Pritchard's words, "The coloring beneath the ocean is all in the lowest tones, merging from deep indigo and purple into the higher, delicate tints of pale greens, grays, and yellows."[83] This palette appears in Pritchard's *Parrot Fish and Poisson d'Or*

FIG. 1.6. Walter Howlison "Zarh" Pritchard, *Parrot Fish and Poisson d'Or amongst the Coral in the Lagoon of Papara Tahiti* (ca. 1910), oil on leather. Courtesy of the Scripps Institution of Oceanography, UC San Diego. According to Pritchard, at a depth of about thirty feet, where he painted this fish (a yellowtail coris), the unaided human eye perceived coloring in the "higher, delicate tints of pale greens, grays and yellows."

amongst the Coral in the Lagoon of Papara Tahiti (1910), donated to the Scripps Institution of Oceanography in 1917, and still on display there today (fig. 1.6).[84] On the backs of his paintings, Pritchard would include specifics about location, subject, and sometimes even the time of day when an image was painted, and he situated *Parrot Fish and Poisson d'Or* at a depth of about thirty feet.

FIG. 1.7. François Libert, "Yellowtail Coris, Terminal Phase—*Coris gaimard*," photograph. When viewed with the help of artificial lighting, the coris's bright colors dazzle undersea, as in this contemporary photo of a coris at the same stage of development as the coris painted by Pritchard.

In the chapter, "No Man's Land Five Fathoms Down," Beebe wrote that at thirty feet, "the harshest, most gaudy parrotfish resolves into the delicacy of an old Chinese print, an age-mellowed-tapestry,"[85] and such is the case of Pritchard's "parrot fish," or, to use the name given to the species today, a yellowtail coris. Seen through air, the coris is brightly colored, and the coloration comes through in a contemporary photograph, which overcomes the underwater haze and renders subjects with brightness and clarity (fig. 1.7).

While Pritchard is true to underwater color at about thirty feet in *Parrot Fish and Poisson d'Or*, the painting imports conventions from the theater for orienting viewers in underwater space. Pritchard started his career in the 1890s designing for the stage. By his account, he caught the attention of Sarah Bernhardt when he went backstage after a production to explain how a costume for an underwater scene should be altered to express submarine perception.[86] Pritchard brought a theatrical sensibility to his underwater settings, evinced in the arrangement of coral reefs in *Parrot Fish and Poisson d'Or*. Here, coral formations function like flats on a stage, to create a sense of recession. The reef in the left foreground, taking up almost three-quarters of the left edge of the painting, demarcates a plane close to the viewer, in contrast to the stalks of living coral extending diagonally to the right, rendered in a pale undersea palette.

Giving the viewer the ability to differentiate the foreground and the middle ground of the scene using reefs as flats, Pritchard takes us into the background of the image by blurring and dimming color. He spoke lyrically of looking off into the underwater haze where "[n]owhere does substance appear beyond the middle distance. Material forms insensibly vanish into the veils of surrounding color."[87] We see such "insensibl[e] vanish[ing]" in how he renders the three yellow butterfly fish in *Parrot Fish*, dimming gradually into the "higher, delicate tints of pale greens, grays, and yellows." This insensible vanishing contrasts with Ransonnet's demarcated zones of green and brown separating the foreground from the middle ground of the coral in his pictures from the diving bell. To such representation, Pritchard adds a familiar compositional technique in painting, threading a diagonal organization running from the bottom left to the top right of the painting to join together foreground, middle ground and background. This diagonal, joining the dorsal fin of the coris, the leafy fringes of the coral, and then the arc of the butterfly fish, is slightly curved, conveying a sense of movement. Another area of the painting suggesting curvature is a swathe on the lower right of blue, shaded with greyish yellow, behind the coris. The information in this swathe is ambiguous. Is it blue coral, like the coral to the left, which would place it in the foreground of the painting? However, the swathe has no fronds or other outcroppings. Is it the aquatic atmosphere? If so, it must represent a gap between the corals, whose arch we can't quite make out, taking us into the painting's background. Given the care Pritchard has used in rendering the subtle coloring of the blue swathe, its placement seems intentional, so its ambiguous hovering between background and foreground might be read as an expression of aquatic perspective and disorientation in the depths.

All these qualities—the muted underwater color spectrum, the theatrical organization of space, the rendition of depth through insensible vanishings, and the curved diagonals organizing the image—are characteristic of Pritchard's aquatic perspective. We see such techniques, for example, in the painting *Moorish Idols*, also donated by Ellen Scripps to the Scripps Institution of Oceanography, similarly painted around 1910. Pritchard also renders these fish, found in the waters off Tahiti, at a depth of thirty feet, and hence the coloring is similar to *Parrot Fish and Poisson d'Or*. Along with using the reefs as flats, Pritchard takes from proscenium theater the technique of exaggerating and compressing the geometry of linear perspective to create depth in a shallow space. The bands on the body of the fish to the left are larger than those of the fish to the right, and as they diminish proportionally, they take the eye back into the scene. This sense of recession is accentuated by optical effects specific to submarine viewing, as the slender crest that terminates the dorsal fin of the fish to the right blurs into the haze. The tall, green coral fronds blur and dim as well, from the more vivid fronds in the middle ground to the ghostly outlines at the back of the scene.

If a vague sense of spatial disorientation characterizes Pritchard's aquatic perspective, the creatures in these paintings are disorienting as well, as they at once are fish and yet have faintly human traits. Thus, in *Parrot Fish and Poisson d'Or*, Pritchard renders the coris to convey the fine details of fish biology and yet also gives his coris an eye more like a human than a fish, with its blue pupil in a white iris. Note, too, the lips, so "distinctly offset from the head coloration," as ichthyologist Philip Hastings observes, which is the case of human lips, in contrast to the lips of the coris as shown in the photo.[88] Indeed, with their bow form brought out by muted red, it is almost as if the fish is wearing lipstick.

Since the Victorian era, biologists had been fascinated by the ways in which marine animals disrupted conventional gender norms, including in their mating, in the distribution of reproductive labor, and even in their sexual identity, as some marine animals are hermaphroditic.[89] It is in keeping with Pritchard's fantasy of the coris as a fish exceeding biological boundaries that the coris can change sex in the course of a lifetime, although I doubt that biologists, let alone Pritchard, knew about hermaphroditism of the coris in 1910. Corises "are hermaphroditic, spawning first as females and changing sex to male, signaled by a significant change in color."[90] The specimen in Pritchard's picture with lips outlined as if with lipstick has the coloring of "a large male." Further, the coris has a phallic appendage—the raised front dorsal fin—accompanying its red-bowed lips.

The faces of the Moorish idols in Pritchard's painting of them blur the boundary between human and fish as well. While they conform to fish biology, their eyes seem to return each other's gaze, with the left fish turning to heed the fish on the right as they socialize companionably in the encircling forms of the reef. Could the companionship of these two fish—who are known to mate for life—be a portrayal of intimacy based on likeness rather than difference? Perhaps Pritchard uses the undersea as a setting to express a queer sensibility that has drawn writers and visual artists to marine life from the nineteenth-century fin de siècle to our present. No scholar has delved into

Pritchard's biography, but the little information available suggests he was gay. He was part of the circle of aesthetes around Bernhardt when he was young, he never married, and he had a close relationship with a young man who lived with him at his ranch in Bishop later in life, until residents broke up the relationship and forced Pritchard to leave his longtime home.

Describing "the metaphysics of light of pagan and Christian neoplatonism" in early Christian art, Panofsky observes how the world, in theology, as in image, is "robbed of its solidity and rationality"; space becomes "a homogeneous and . . . homogenizing fluid, immeasurable and indeed dimensionless."[91] For Panofsky, the system of linear perspective invented during the Renaissance affirms the presence of the human over the divine, in an episteme he terms "anthropocracy." The emphasis on the human in linear perspective has plural implications, from abstraction—"perspective mathematizes this visual space"—to human experience—"and yet it is very much *visual* space that it mathematizes."[92] Building on Panofsky, among others, Martin Jay calls this version of perspective that fuses abstraction and realism "Cartesian perspectivalism." It belongs to a version of humanism that would prevail in western Europe over subsequent centuries, where vision among the senses becomes the ultimate expression of knowledge and domination.[93]

Aquatic perspective intervenes at the other end of this history, revealing a realm where "anthropocracy" is ill at ease. As I have described, Ransonnet's and Pritchard's underwater imagery transports us to a realm that unsettles our terrestrial points of orientation, from our sense of depth to our understanding of species boundaries. Further, the atmosphere itself starts to take on substance, turning into an element that can be formed, at once material and elusively fluid.

Tellingly, in the imagery of Ransonnet, and in almost all of Pritchard's art, there is no diver in the scene. Rather than locating the human in the picture, the submarine realm becomes, to cite Elizabeth DeLoughrey an "uncanny medium that distorts our terrestrial-bound understanding of figures, time and space."[94] DeLoughrey formulates the undersea uncanny while writing about the contemporary underwater installations of British artists Jason deCaires Taylor, yet her comments help make sense of Ransonnet's engravings and Pritchard's paintings as well. Both Ransonnet and Pritchard are at the beginning of an artistic lineage that today is a frontier of environmental art. Thus, underwater art museums, inspired and in some cases overseen by deCaires Taylor are opening around the world.[95]

DeLoughrey assimilates the undersea uncanny to her concept of "new sea ontologies." In pondering how humanism is displaced by the human encounter with the realities of water, Stacy Alaimo has proposed the concept of "transcorporeality," which, in her words, describes a condition where the "material self cannot be disentangled from networks that are simultaneously economic, political, cultural, scientific, and substantial."[96] Beebe looked to avant-garde representations in characterizing the "architecture and design of undersea" that he found Pritchard to render so expressively, writing that if it "were copied closely," "[no] change need be made in the most weird, most ultramodern of ballets."[97] Yet some of Pritchard's contemporaries admired his art for its more traditionally spiritual rather than avant-garde perspective. When Jiro Harada, His Imperial Japanese Majesty's Commissioner to the Panama-Pacific International Exhibition, described Louvre curator Jean Guiffrey's reaction to Pritchard's work in a letter to Pritchard written from San Francisco in 1915, Harada said that Guiffrey used "a remark in French which you told me meant 'words cannot express it.'" In Harada's assessment, Guiffrey was "deeply moved . . . by your paintings, their uniqueness, charm and delight in the mysterious harmony of colors revealing something more than the eye can see." "Wonderful," Harada reports Guiffrey commenting about "the submarine picture of shoals of fishes apparently worshipping the image of Buddha."[98]

Pritchard referenced theology as well, Christian rather than Buddhist, in recounting his first dive with a helmet in Tahiti that took him over "the Passage of Papeete, over six hundred feet deep," in an interview with the magazine *Asia*.[99] "When one leaves the resounding, splashing surface . . . to enter the depths below, one is astounded by the sensation that some sudden magic spell has swept away every drop of the water." Comparing breath-hold diving to helmet diving, he noted that a phrase from Cardinal Newman's "Dream of Gerontius," describing "the experience of the spirit immediately after death," which had floated through his mind while breath-holding, became actualized: "As though I were a sphere, and capable / To be accosted thus."[100] During this dive, Pritchard "felt absolutely happy," and his memory of it mixes evocations of ecstasy and dream with close environmental observation, such as when he notes, for example, "I was an atom suspended in the center of a limitless, horizontal-banded sphere of translucent color."

In the decades when Pritchard was painting, film and photography were beginning to exhibit the undersea. When Williamson's documentary first screened in 1914, "[t]he Washington papers cried: 'Films that pierced the sea—each picture an absolute revelation.'"[101] Exclamations of wonder accompanied spectators' reactions to submarine film, as visitors had marveled at the aquarium, and indeed such exclamations resound each time we access a new aspect of the submarine realm. Yet these exclamations are the evidence, not the explanation. All the fantasies considered in the subsequent pages explore new possibilities for aesthetics, knowledge, and emotions prompted by the underwater realm's displacement of human mastery, which I have in this section called aquatic perspective as symbolic form. In contrast to creating aquatic perspective with traditional media, like Ransonnet and Pritchard, submarine filmmakers expand our visual imagination, drawing on the capacities and conventions of the twentieth century's world-changing new exhibition medium of film.

1.3. The Early Underwater Cinema of Attractions

When film was first developed in the late 1890s, there was sufficient public interest in the underwater environment for directors to long to show it, even before the Williamson photosphere enabled its capture. Jonathan Crylen has noted as early as 1898 films by Georges Méliès, including the *"Visite sous-marine du Maine* (1898), one of several films that year to feature the USS *Maine* shipwreck; and such *féeries* [fantasy cinema inspired by fantasy theater of the later nineteenth century] as *Le Voyage dans la lune* (1902), *Le Royaume des fées* (1903), *Le Voyage à travers l'impossible* (1904), and *La sirène* (1904)." Crylen also notes that "Pathé's *Un Drame au fond de la mer* (1901), a reworking of a contemporary British stage melodrama, also featured an undersea set."[102] Given the year, the melodrama was presumably the 1897 *White Heather.*

Submarine films before underwater filmmaking used a variety of media to express the undersea. In *A Drama beneath the Sea* (*Un drame au fond de la mer*), a "large scale cyclorama painting" was the submarine backdrop.[103] In the *Submarine Visit to the Maine* (*Visite sous-marine du Maine*), Méliès took spectators underwater by using an aquarium to provide the movement of water and fish, whether he shot the action scenes through a "thin fish tank in front of his actors" or used superimposition.[104] In *The Realm of the Fairies* (*Le royaume des fées*), Méliès also submerged miniature ships in tanks, capturing air bubbles rising from them.

One aspect of submarine materiality that resisted Méliès's exhilarating creativity, however, was movement. When a diver climbs up the ladder to the *Maine* in a *Submarine Visit to the Maine*, he is visibly straining against gravity, in contrast to the glide of underwater ascent. In *The Realm of the Fairies*, the dancers swim horizontally, but their movements show little resistance, notably their legs, which flutter up and down. The film includes a parade of wonderfully whimsical undersea inhabitants that glide smoothly, but the glide does not undulate, instead proceeding in a straight line from left to right, presumably on tracks. When a swordfish in this parade stops, like a stubborn horse, his human rider whips him with a sword cutting through the air.

The use of special effects to simulate the marine environment was superseded in 1914 with Williamson's photosphere. Sitting in the submersible encased in air, the camera person and the camera looked out into the deep. The photosphere could operate at different depths thanks to a flexible tube and was connected by a telephone to the surface, enabling the camera operator to receive instructions.

The first underwater footage shot by Williamson from his photosphere premiered for general audiences at the Smithsonian Institution on July 16, 1914. This screening was filled to capacity, as was a second.[105] In the audience was marine scientist Charles Townsend, director of the New York Aquarium, which had hosted a first exhibition of Pritchard's art in 1913 and would host a second in 1916. With Williamson's film, Townsend and a number of scientists at the screening "were looking at sea life in its natural habitat for the first time in their careers."[106] So revelatory was the film that Townsend "was even finding species he had never before recorded."[107] Initial screenings were at scientific institutions, such as the first screening in New York at the American Museum of Natural History, attended by Albert I of Monaco, a collector of Pritchard's and Ransonnet's paintings. The film was first named *The Williamson Submarine Expedition*. For commercial release, it would be retitled *The Terrors of the Deep*, then *At the Bottom of the Ocean*, and eventually *Thirty Leagues under the Sea*. When "the film went on tour . . . [it] garnered rave reviews the world over."[108]

Only one reel of the five-reel *The Terrors of the Deep* is easily available, disseminated online by the Nederlands Film Museum.[109] However, audiences today do have online access to the 1916 film *Twenty Thousand Leagues under the Sea*, directed by Stuart Paton, with extended sequences using footage by Williamson, shot undersea in the Bahamas. This film coincidentally premiered in Chicago the day after a German U-Boat, "U-53 suddenly showed up near the Nantucket Lightship, and in full accordance with international law and America's neutrality, began ravaging the sea lanes into New York Harbor."[110] As Burgess details, "the premiere night audiences offered generous rounds of applause on seeing the undersea segments of the film," and in "eight weeks' time, the Williamson-Universal production had completely paid for itself."[111] Universal held off the picture's New York premier until the Christmas holidays. When it was released, it "went on to enjoy a two and a half month booking; the second longest run of any picture ever shown in New York up to that time."[112] Throughout 1917, *Twenty Thousand Leagues under the Sea* toured across the United States, screening to packed theater houses. After World War I ended, the film would "cross the Atlantic for an enviable distribution—especially in Jules Verne's native France."[113] The topicality of submarine warfare was one reason for the film's resounding success, but above all, audiences were thrilled to see a new realm of our planet. A review in *Motion Picture News* from January 6, 1917, is expressive: "Whole gardens of undersea life hold and fascinate and seem like a dream. . . . Some will say, 'impossible,' 'ridiculous,' but the majority will say 'marvelous,' how did they do it?"[114]

In this section, I consider how Williamson organized his underwater imagery to transform the submarine realm from physical environment into underwater film set. My comments are brief because Burgess and Adamowsky discuss Williamson's success, and Crylen devotes an entire chapter of "The Cinematic Aquarium" to the photosphere's exhibition of the undersea. I am guided by Crylen's astute insights that Williamson organizes the submarine environment as an attraction, following early cinema theorist Tom Gunning's concept to designate a popular type of early cinema that celebrates the medium's ability "to *show* something."[115] Crylen's discussion also supports my suggestion that one challenge faced by underwater filmmakers is to organize the underwater environment's features so that they make sense to terrestrial eyes. Crylen connects Williamson's organization of the undersea to preexisting exhibition formats, notably the aquarium and the panorama. The photosphere orients its views with the rectilinear format of the aquarium, even if the edges of the circular tube of the photosphere's viewing window are visible in the corners of some shots. As inspiration for the "steady, lateral movement" of how underwater scenes unfold, Crylen cites both the "panorama on rollers and an ambulatory [spectator's] gaze before such a scene."[116]

Gunning relates the cinema of attractions to other popular spectacles of the time, such as vaudeville and the carnival side show, where indeed brief films were sometimes projected. Gunning contrasts the cinema of attractions with narrative cinema. By 1907, he comments, the cinema of attractions started to wane in popularity, although it "does not disappear with the dominance of narrative, but rather goes underground, both into certain avant-garde practices and as a component of narrative films."[117] Among narrative films, Gunning mentions the genre of the musical, where the cinema of attractions continues, as narrative comes to a halt for the pleasure of song and dance. When Williamson takes the cinema of attractions below, he adds the exhibition of a novel planetary environment to the cinema of attraction's repertoire.

We can see the grip of such environmental exhibition in the underwater sequences shot by Williamson for Paton's *Twenty Thousand Leagues*, which is on balance a narrative film. The novel follows the suspenseful adventures of Nemo's fantasy submarine and its captives, and the 1916 film, too, has sequences of underwater suspense such as his divers' encounters with ferocious predators. Yet the first extended underwater sequence in this film is a panorama over the seafloor, exhibited as an object of viewing pleasure. In this remarkable sequence, which lasts about eight minutes, Nemo invites his captives to look out of his "magic window" and revel in "the marvelous mysteries of the deep" (ca. 29:55–38:00). The magic window presents the underwater environment with the orientation of an aquarium tank—yet reveals scenes no aquarium ever contained. The film intercuts these views with views of characters inside the submarine expressing astonishment, and with dialogue cards where Nemo explains what they are watching, in the didactic spirt of Verne's professor narrator Aronnax. Thus, the titles point out coral formations, "skeletons of the little coral polyp—a marine animal," or "the wreck of an old blockade runner" (ca. 34:21, 34:58).

Movement is part of the fascination of vaudeville, which Gunning sees as shaping the cinema of attractions. Although aquarium tanks could have incorporated helmet divers, these nineteenth-century underwater gardens were sealed off from human presence. Watching humans move underwater in film, spectators reveled in their glide and their acrobatics in an atmosphere where they are released from gravity—a novel view to all but a few professional divers. Further, they enjoyed the movements of sea creatures, notably animals that aquariums did not yet have the technology to maintain in captivity, such as sharks.

Even before underwater film, underwater dance was a type of vaudeville exhibition, made famous by the Australian swimmer Annette Kellerman. Kellerman starred in vaudeville performances in the early twentieth century, where "[s]he developed an imaginative stage act with which she topped the bill at the London Hippodrome, swimming and diving in a huge glass-fronted tank."[118] Kellerman was a living connection between vaudeville and dance in underwater film, when this accomplished waterwoman pivoted to film at the

FIG. 1.8. James R. Sullivan, *Venus of the South Seas*, ca. 24:57. Annette Kellerman holds a hand mirror, her image duplicated in a mirror behind her, recalling the illusions of vaudeville, as she takes the cinema of attractions into the depths.

time of Paton's *Twenty Thousand Leagues*. Kellerman, who would go on to be the subject of *Million Dollar Mermaid*, the 1952 biopic loosely based on her life starring Esther Williams, first appeared in the medium of film performing submerged in *A Daughter of the Gods*, also released in 1916.[119] This film was shot "at Montego Bay in Jamaica with a cast of 20,000 and a budget of more than one million dollars, making it the most expensive film to that date. The principal set was an 8.1-hectare Moorish city, which was burnt to the ground in the final scenes; as well there was an intricate underwater 'mermaid village,' shot by a camera encased in a specially designed diving bell. The waterfall scene was the first in movie history to be done completely in the nude."[120]

Kellerman would go on to make a number of aquatic films, although "only *Venus of the South Seas* has survived complete."[121] Underwater scenes in this film stop its narrative, such as it is, and create scenes of attraction beneath the sea. Kellerman glides and dances, and moreover, these scenes thematize the pleasure of looking. In vaudeville, Kellerman performed with "mirrors [placed] in strategic positions around her tank."[122] *Venus of the South Seas* takes us down to the underwater dressing chamber of a mermaid, which features both a handheld mirror, and a mirror as backdrop. Kellerman visits the dressing chamber in her roles both of the mermaid and a little princess who pays the mermaid a visit, and in both roles, she admires herself in these looking glasses, while the backdrop mirror duplicates her image (fig. 1.8). Scenes of such admiration intersperse with the mermaid's underwater ballet and the princess exploring, walking through water with the comfort, but not the gait, of walking on land. Throughout the underwater scenes, whether as little princess or mermaid, Kellerman has no breathing aid, and keeps her eyes open as well, showing the ease in the water of this extraordinary athlete who was able to breath-hold for over three minutes. Since the film was projected in color, one can only imagine the magnificence of its underwater views from the black-and-white print available online.

Until now, I have emphasized the obstacles for rendering scenes created by the underwater haze. Yet *Venus of the South Seas*, as well as *Twenty Thousand Leagues*, shows it as an enhancement to the pleasure of underwater looking. In *Venus of the South Seas*, the haze softens the contours of the body and creates mysterious allure. The mystery is beautiful when Kellerman dances; it is menacing when the shrouded creatures are sharks. The underwater haze also can create a tone of gravitas, as at the film's ending. As in Verne's sequel to *Twenty Thousand Leagues*, *The Mysterious Island*, Paton's *Twenty Thousand Leagues* concludes with Nemo's death and the *Nautilus*'s disappearance beneath the waves, lost to humankind (ca. 1:36:44–1:40), in two separate sequences, first showing Nemo's burial, and then the scuttling of *Nautilus*. The haze of the burial is mournful, while the haze of the scuttling scene is more enigmatic, holding out the promise that the fantastic ship may still lie somewhere on the sea floor awaiting rediscovery.

In more theatrical fashion, the underwater haze can also help mask artifice. In Paton's film, the murky underwater atmosphere helped naturalize a mechanical octopus designed by Williamson. This octopus was the first of many mechanical legendary predators in underwater films, in a lineage including the giant squid in the 1954 *20,000 Leagues* and Bruce, the shark in *Jaws*. In all these films, the aquatic atmosphere blurs the forms and cloaks mechanics—the 1954 *20,000 Leagues* used night and a topside storm as well.

The octopus, like the sharks, is an attraction of Paton's *Twenty Thousand Leagues* and, indeed, yet another connection inspired by biology (even if evoked by artifice) to vaudeville, which also included the exhibition of dangerous predators, such as lions and bears. The sharks are a particularly captivating element of *Twenty Thousand Leagues*, and they star in the underwater hunt through the forest of Crespo, which unfolds over an eight-minute sequence, lengthy by the standards of underwater film throughout its history, from the first view of the submarine's hatch opening at approximately 45:38 to the moment when it closes at about 53:53. Paton does structure this sequence as a

FIG. 1.9. Stuart Paton, *Twenty Thousand Leagues under the Sea*, ca. 51:48. Divers and large sharks appear together in a thrilling sequence that may be the first view of these apex predators *mobilis in mobili*, in their element.

rudimentary adventure: an exploration concluding with a shark attack. Just as, if not more, compelling are vistas of sharks gliding. Another attraction is the presence of divers and sharks together (fig. 1.9), a presence that Williamson featured as well in his documentary *The Terrors of the Deep*. The observations of critic André Bazin about such exhibition of actual danger explain the power of the scene. Bazin contrasts representations putting predator and human in the same frame with the attenuated impact of danger created through montage, such as the alligator sequence in filmmaker Robert Flaherty's *Louisiana Story* (1948).

In another episode of danger, the encounter of a free diver and the mechanical octopus, the film draws suspense from a powerful narrative poetics. Visual pleasure meshes with a storyline that will become a mainstay of underwater narrative, in both film and TV, which is the dangers of the environment's toxic atmosphere for humans. In the octopus attack on the free diver, in *Twenty Thousand Leagues*, we see the diver struggling while entwined by the octopus, intercut with views from the submarine of his battle and views of Nemo, who dons helmet diving gear and goes to his rescue. The views of the diver in this sequence look as if they were captured in a single take. As a result, they intensify the suspense with questions about how long the diver can hold his breath. The screen time of the episode, which is two minutes, offers a plausible timeframe were the diver actually to have acted the entire sequence underwater in one take (the diver jumps in ca. 59:48; the octopus entangles the free diver around 1:00:12; and he is taken up to the surface within a minute and a half—ca. 1:00:21). The view of an actor, as if in a single take, submerged and thwarted from surfacing, will go on to create suspense to across a century of underwater adventure. As I will subsequently discuss in part 2's analysis of *Sea Hunt*, this suspense formula dates back to Verne's novel and its chapter "Not Enough Air." We will see it used repeatedly in *Sea Hunt* (1958), and one recent memorable variation is Tom Cruise's free dive to swap files in the 2015 *Mission: Impossible—Rogue Nation*.[123]

1.4. Surrealism's Aquatic Perspective: From *Piscinéma* to Aquatic Humanism

Among the types of film in which the cinema of attractions remains alive, Gunning includes works of the avant-garde. Surrealism illustrates this continuity, offering more unconventional pleasures of looking than vaudeville to publics seeking new optical and perceptual experiences. In 1925, the year after surrealism proclaimed itself a movement, its leader, André Breton, saw a Williamson documentary, screening in France with the title *The Wonders of the Sea* (*Les merveilles de la mer*), and was enthralled. As he wrote in a letter to his wife at the time, Simone Kahn: "[T]he walk of the divers against the currents— that's ballet. And then the octopus is so beautiful."[124] Sean O'Hanlan follows Ann Elias in emphasizing how Breton, like other surrealists, marveled at the new vistas revealed when film went below, and considered underwater vistas to effect *dépaysement*, which in surrealist terms was unsettling, displacing what the surrealists called "bourgeois" conventions. By "bourgeois," surrealists meant a loose affiliation of artistic techniques, which they also deemed realist and traditional. These techniques included Panofsky's linear perspective; nineteenth-century realism, in literature, as in art; positivist epistemologies; Cartesian rationality; and bourgeois politics. *Dépaysement* is a fortuitous term for a study on underwater aesthetics, as it of course also contains the notion of land— *pays*—that is being in some way displaced or undone. I will translate it in this section as "displacing," because place implies situation and also thinking of the buoyant quality of water that displaces the weight of the body submerged in it.

In connecting underwater worlds to surrealist displacement, critics have notably examined the oeuvre of marine biologist and surrealist Jean Painlevé. As Steven Erickson writes, "[I]t's worth noting his films' insistence on denaturalizing conventional ideas of gender and sexual orientation. *The Sea Horse* (1934) shows a male seahorse giving birth, while other shorts depict hermaphroditic animals and asexual methods of reproduction; although subtly and gently expressed, there's a queer sensibility at work."[125] For Painlevé, a species in which the male carries the eggs and gives birth makes it a "symbol of tenacity [that] joins the most virile efforts with the most maternal care," which he offers to "those who wish for a companion who would forgo the usual selfishness in order to share their pains as well as joys."[126] Furthermore, the appearance of these animals challenges us to assign them a gender, as both male and female sea horses move with conventionally feminine delicacy.

Given the existing studies on Painlevé, I will not explore further his fascination with the underwater realm. Rather, I focus in this section on the interest of surrealist leader André Breton and premier surrealist filmmaker Man Ray in underwater imagery, and notably seeing through water, as techniques of surrealist displacement. Elias cites an intriguing passage from Breton in his "Surrealism and Painting," "where Breton brought to mind 'the Marvels of the sea a hundred feet deep,' [writing] that only the 'wild eye' freed from habit can be fully receptive to the magical sensations of the outer limits of the world."[127] O'Hanlan connects the "wild eye" freed from habit to submarine materiality: "It was precisely this disorientation of habitual perception and salvage of attention that Breton sought as he referred to the physical nature of vision underwater in explicating Surrealism's perceptual experience in unfettered thought. . . . The 'marvels of the sea' were not only metaphors for this new 'scale' of vision, but a literal reference to both the limitations and radical possibility of 'penetrating sight' in a *dépaysé* environment."[128]

Breton includes a still from the Williamson photosphere among the photos illustrating his volume of essays *Mad Love* (*L'amour fou*) (1937). As Elias has noted, Breton identified the corals as those of the Great Barrier Reef—what he called "the treasure bridge of the Australian 'Great Barrier' Reef"—although the Williamson still featured corals in the Bahamas.[129] The photos that punctuate the essays in *Mad Love* explore different aesthetics of surrealist displacement, which were, for Breton, like other surrealists, fostering perception that would bring about the surrealist revolution when dream and action merge. When Jean Cocteau described his procedure in filming his first version of *Le sang d'un poète*, he likened his efforts to wrest poetic images from the unconscious to underwater filming. He wrote, "in *Blood of a Poet*, I try to make [*tourner*] poetry like the Williamson brothers make films [*tourner*] under the sea. It was a

Brassaï

FIG. 1.10. Brassaï, "Between the Hedges of Blue Titmouses of Aragonite and the Australian 'Great Barrier Reef.'" Composite image used to illustrate André Breton's essay "Convulsive Beauty" when it appeared in *Minotaure*, no. 5 (May 12, 1934): 11. Brassaï's crafted photographs of coral skeletons create a sense of surrealist displacement recalling underwater optics in the atmosphere of air.

question of lowering in myself the bell that they lower into the sea, to great depths."[130]

In *Mad Love*, Breton used the still from Williamson's photosphere in the context of describing what Breton called "convulsive beauty," a term he devised for an aesthetic effect, in whatever medium, often signaled by a shiver, that dissolved categorical oppositions essential to Western thought, such as the opposition between life/death, real/imagined, and high/low. The oppositions he paired to illustrate the category-crossing of convulsive beauty were themselves unexpected: convulsive beauty will be "veiled-erotic, fixed-explosive, magic-circumstantial."[131] Breton saw coral as a physical expression of convulsive beauty because it was at the juncture of the mineral and living, as well as rock and water. To cite his wording, in characteristically difficult style, "those absolute bouquets formed in the depths by the alcyonaria, the madrepores," are phenomena "[w]here the inanimate is so close to the animate that the imagination is free to play infinitely with these apparently mineral forms, reproducing their procedure of recognizing a nest, a cluster drawn from a petrifying fountain."[132] For Breton, water expressed a sense of presence out of history; earlier in *Mad Love*, Breton wrote, "To glide like water into pure sparkle—for that we would have to have lost the notion of time."[133] Coral, in its petrification, had a transcendent quality as well for Breton, which he also found in crystal, but in the natural environment of the coral reef, crystal's "omnipotent reality," disappeared, "reintegrated, as it should be in life, into the dazzling sparkle of the sea."[134]

When Breton first published his essay on convulsive beauty in the surrealist art journal *Minotaure* in 1934, it was accompanied by a composite page with photos of crystals and coral skeletons, exquisitely crafted by photographer Brassaï, whose photos are featured throughout *Mad Love*. Brassaï created a sense that the corals were underwater through such techniques as close-ups, cropping, and erasure of background information (fig. 1.10). When Breton republished the essay in *Mad Love*, he substituted a view of the corals from the Williamson photosphere that he found in the *New York Times* (fig. 1.11). Breton would seem to agree with his contemporary, the scientist adventurer William Beebe, that undersea one observes a physical reality comparable to what Beebe called "ultramodern" aesthetics. Indeed, views of the underwater realm recall a range of surrealist visual effects. The ambiguous identity of marine life, the impact of the underwater haze on linear perspective, and the unfamiliar angles so easy to achieve moving in its three dimensions all enable presentations of phenomena as quite strange. What I have called aquatic perspective for Ransonnet and

FIG. 1.11. Still image of corals from J. E. Williamson's photosphere. The text on the back of the photo reads, "Expedition océanographique J. E. Williamson dans les profondeurs de la mer." World Wide Photo. Courtesy of the Association Atelier André Breton. When Breton published "Convulsive Beauty" as the first chapter of *Mad Love*, he substituted this photograph from Williamson's photosphere for *Minotaure*'s composite image: a view of the actual submarine environment would seem to outdo fine artistry.

Pritchard in its surrealist iteration matches not only the movement's general project of dislocation and Breton's convulsive beauty but also an aesthetic that surrealist philosopher Georges Bataille called the *informe*. As Rosalind Krauss explains, Bataille applied this notion to disturbing physical representations intimating a mental state characterized by "the removal of all those boundaries by which concepts organize reality, dividing it up into little packages of sense."[135]

Among the photos in *Mad Love* that use water to foster convulsive beauty, Breton included a picture of dancer Jacqueline Lamba (who would become his second wife) submerged in water, with the caption "Seeming to swim."[136] The photo of Lamba comes from a series that Rogi André, an alias for the photographer Rosza Klein, took in a tank in 1934. In his text, Breton compared Lamba to Undine, and the photo chosen for *Mad Love* shows her as part human, part water nymph, poised in a supple water-efficient glide, with her eyes open and her countenance relaxed, her nudity discretely veiled by the haze that enhances her glamour. Other images of Lamba in André's series disturb her human appearance in a fashion more resembling Bataille's *informe*. Thus, in one, André uses the obscure atmosphere of water, lighting, and "lenses specially modified by André Kertesz and . . . a mirror," to turn Lamba into a strange interspecies creature.[137] Pritchard's species-crossing fish are charming compared with Lamba, shown with an attractive female lower body attached through a hyperflexible torso to a blurry head that at first glance lacks human features, and to bony arms, where a bent wrist can also be read as a leg joint of an exoskeleton.

Man Ray was the filmmaker among the surrealists who was most fascinated with the opportunities offered by the optical properties of seeing through water to further surrealist displacement. He first explored such potential in *The Starfish* (*L'étoile de mer*) (1928), whose subject references the creature as evoked in an eponymous poem by surrealist poet Robert Desnos. Man Ray's film exhibits the starfish as an enigmatic bearer of meaning—an animal/object—shown as a specimen, both dried, and in a glass container, as a paperweight.[138] As in some of Painlevé's films, the starfish's sexuality is disturbing, with its ambiguously phallic limbs recalling the legs of the female protagonist in the opening sequence of the film that both fascinate and repel the male protagonist. Further, the starfish suggests the Freudian notion of the *vagina dentata*, a vagina with teeth, and in the footage underwater, the starfish has wrapped its legs around prey that film critic P. Adams Sitney identifies as a sea urchin. Man Ray brings his perturbing vision of a starfish to life with underwater footage, showing it bristling with undulating arms. However, Man Ray did not take his camera underwater, instead using footage provided by Painlevé.

Yet it is important not to overlook the film's attention to seeing through water, along with its more evident representation of the surrealist interest in perturbing sexuality. "To watch with something different than our everyday eyes," to cite filmmaker Jean Vigo (subsequently discussed), is a trope of surrealist film from Luis Buñuel and Salvador Dalí's shocking image of an eye sliced with a razor at the opening to *An Andalusian Dog* (*Un chien andalou*) (1929).[139] *The Starfish* invites us to look with a marine eye in its initial montage, which starts with a rotating starfish placed in a circle like the Vitruvian man (fig. 1.12). After the title of the film, the montage includes views of a window that slowly opens into its world, a conventional metaphor in a film or photo for the invitation to look, which will be reinforced as the window closes at the

FIG. 1.12. Man Ray, *The Starfish*, ca. 0:52. The arms of the starfish captured in a circle evoke and transform the Vitruvian man.

film's end. The window that opens in *The Starfish* is made of the stippled glass that Man Ray will use as one of his lenses shooting the film. This "stippled lens . . . distorts many of the images" as Sitney notes, in a film that excels in "presenting a still life in fluid contours."[140] Further, the oval shape of the window recalls the portholes of boats, thus taking us out to sea, and Man Ray blurs everyday scenes such as a couple walking along a road, as if seen through water (fig. 1.13). When the film then does transport spectators waterside, it moves from crisply outlined boats to "eerie silhouettes" in the marine fog.[141]

Many shots of otherwise legible places and character interactions in the film are out of focus. Along with stippled lenses and fog, Man Ray created the film's blurry focus using sheets of gelatin.[142] The blur allowed Man Ray to show his female character nude, and escape censorship in the opening sequence, yet it persists throughout the film. Sitney connects the blurred focus to the abiding avant-garde preoccupation with ways of seeing and the techniques of art.[143] As he explains, Man Ray alternates shots in and out of focus with different lenses and thus "points first of all to the very fact that *films are shot through lenses. L'Etoile de mer* is a film about seeing the world through layers of glass."[144] Sitney further interprets the variety of techniques used by Man

FIG. 1.13. Man Ray, *The Starfish*, ca. 1:32. Man Ray dissolves the familiar contours of a couple, developing his film's invitation to look with a marine eye.

Ray to create such blur as reminder of the work of image-making, calling attention to the speed of the camera and "even the atmosphere through which it gathers its light."[145] Sitney's comments about optical technologies in *The Starfish* also mesh with an environmental approach to making sense of the film, whose aquatic *dépaysement* extends from marine biology to the potential of underwater optics.

Man Ray would dive deeper into aquatic *dépaysement* in his subsequent film *The Mysteries of the Chateau of Dice* (*Les mystères du Château du Dé*) (1929). He still did not take the camera underwater, but he shot a pool topside as part of an aesthetic that he called in a catchy intertitle "Piscinéma" (ca. 12:14)—fish or pool cinema. In Man Ray's time, critic Robert Aron noted Man Ray's marine interest disapprovingly, when he inveighed against avant-garde abstraction, which he epitomized with *The Mysteries of the Chateau of Dice*: "All the elements are dislocated, the natural kingdoms confused. . . . Man Ray, following his diver, discovers a new universe which opens up through *piscinéma*."[146] The seas are so fascinating and so much about them remains to be discovered, that saltwater subjects are often conflated with underwater aesthetics. However, freshwater and artificially created bodies of water such as a pool remain opportunities for aesthetic innovation across the history of underwater film, evincing the creative potential released by Man Ray.

In *The Mysteries of the Chateau of Dice*, the film's thin narrative, if narrative there is, takes us on a trip from Paris to a mysterious building that is no historical chateau but rather the Villa Noailles in Hyères, on the Riviera, a gem of contemporary architecture completed in 1927, designed by architect Robert Mallet-Stevens for Charles and Marie-Laure de Noailles. Its setting and geometric, rectilinear lines illustrate a type of modernism celebrating clear and distinct vision, and thus the villa is a fitting starting point for surrealist challenges to such a celebration of reason and light. The de Noailles, moreover, were enthusiastic about both avant-garde currents of their era.

In Man Ray's film, the villa is inhabited by a troupe of lithe young male and female actors. They wear stocking-like masks that show they have human countenances, but not their individuality, and that same lack of individuality characterizes their similar body types, anticipating the resemblance of dancers in ballets by George Balanchine. Their dehumanized quality resonates with an intertitle, as the film starts to explore this villa, asking "this question, this human question: Where are we?" (ca. 9:00). The harmony of the villa and the contours of its athletic inhabitants mutate when the film explores the villa's indoor pool. In this scene shooting into, if not through, water, the female athletes at first turn to aquatic beauties that the intertitle compares to minor deities. Their faces remain hidden as they cavort, whether by mask, blur, or keeping them underwater, as in the moments around the intertitle "Underwater Eve" (Eve sous-marine, ca. 11:29). Then, however, shown through water, the lithe bodies grotesquely distort, bloated by the aquatic atmosphere (fig. 1.14).

The most famous surrealist liquefaction of terrestrial reality is the description by Louis Aragon in *The Paris Peasant* (*Le payson de Paris*) (1926), set in an outmoded arcade slated for destruction, the Passage de l'Opéra. There, the shop window reminds Aragon of aquarium tanks and the goods exhibited turn into an "aquatic dreamworld," in an underwater *féerie*, leading Aragon into an extended erotic fantasy featuring commodities and his desire.[147] When the intertitle *piscinéma* appears in *The Mysteries of the Chateau of Dice*, Man Ray offers a different, yet also radical, liquefaction, dissolving the boundary between room and pool. He creates this dissolve, notably through projecting onto the walls of the room the same undulating play of light and shadow visible in the depths of the pool, including shadows of the swimmers (ca. 12:34) (fig. 1.15). The intertitle "Underwater Eve" appearing in the pool sequence

FIG. 1.14. Man Ray, *The Mysteries of the Chateau of Dice*, ca. 11:54. The underwater Eve: Shot through the surface of water, the lithe contours of the swimmer dissolve and distort.

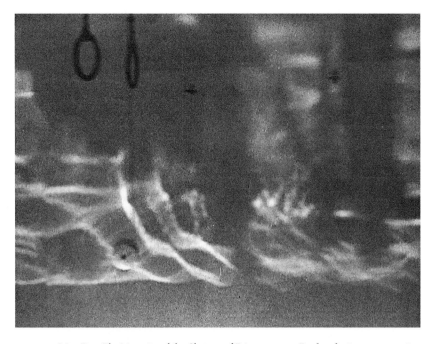

FIG. 1.15. Man Ray, *The Mysteries of the Chateau of Dice*, ca. 12:34. Pool melts into room owing to lighting in Man Ray's *piscinéma*.

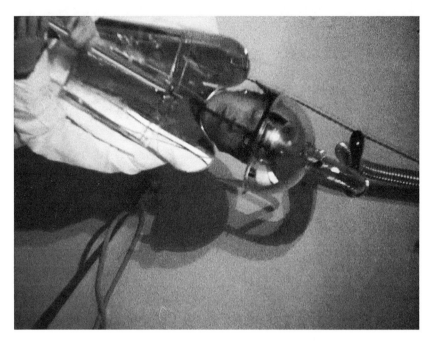

FIG. 1.16. Man Ray, *The Mysteries of the Chateau of Dice,* ca. 14:59. Helmeted Minerva.

echoes the title of a novella by Villiers de l'Isle-Adam, beloved by the surreal-ists, further adding a futuristic cyborg to deities in the film's marine response to its (post)"human question."

The Mysteries of the Chateau of Dice contains one more aquatic fantasy that suggests a new type of submarine knowledge meshing future and past: the figure of "helmeted Minerva" (ca. 14:48), to cite the intertitle, the god-dess of wisdom. When helmeted Minerva appears, the actress is shown in a medium shot (fig. 1.16). She retains the stocking mask of the other char-acters and wears a helmet of shiny chrome, with flaps that run down her cheeks, befitting the classical goddess. Her garb, however, is no longer the revealing workout clothes used by the athletes but rather a loose-fitting blouse that drapes in a manner evocative of a helmet diver's suit. Man Ray suspended helmeted Minerva from a chain, at once referring to the me-chanical context of the cyborg and the tether connecting the helmet diver to the surface. Man Ray then turns attention to the power of the camera and rotates the goddess upside down almost 180 degrees until the camera cuts away. This movement at once implies conventional wisdom turned on its head, the ability of the movie camera to upend conventional views, and three-dimensional movement so easy to achieve in the underwater atmosphere.

When Aron criticized Man Ray's abstraction, he suggested the importance of humanism, which he illustrated with an underwater scene from Buster Keaton's *The Navigator* (1924). In this scene, Keaton dons a diving suit and goes below to fix a jammed propeller, where he performs silent comedy with liquid physics. Aron emphasizes the pathos of a moment when Keaton's air supply is cut off: "In deep sea water, enclosed in a diving suit, near the anchor of a ship, Buster Keaton lifts his arms in supplication. The desperate gesture of a man obliged by social reasons to descend to the sea bottom where he is threatened with asphyxiation, and still wins."[148] Obscured a bit by Aron's pathos, the underwater scene features comedy that is slapstick if not surrealist, as Keaton fends off a swordfish by grabbing another swordfish and using it to duel, before defeating an octopus that is much less elaborate than Williamson's mechanical creature.

Surrealist displacement is reconnected both to human experience and to narrative in *L'Atalante* (1934), the only full-length film by Jean Vigo, who died at twenty-nine of tuberculosis, and who would go on to influence generations of directors. Vigo at once appreciated surrealism and expressed commitments to Marxism and to popular life. He spoke his previously quoted phrase, "[t]o watch with something different than our everyday eyes," during a lecture at the Vieux-Colombier theater in Paris when his short film *A propos de Nice* was screened in 1930.[149] In this lecture, he also expressed aspiration toward a "social cinema" that would address "society and its relationships with individuals and things."[150]

Vigo, who was friends with Painlevé, honed techniques he would return to in *L'Atalante*'s memorable underwater sequence when he made a short documentary about French freestyle champion Jean Taris, *Taris, King of the Water* (*Taris, roi de l'eau*) (1931). Although Boris Kaufman, brother of pathbreaking Soviet director Dziga Vertov is not named in the credits, Kaufman worked on all Vigo's films as cinematographer. In *Taris*, shot in and around a pool, Vigo, like many great creators of underwater cinema, utilized the impact of the difference between topside and underwater views.

The subject of the film is Taris's swimming technique. The first half of the film shows and explains this technique from the surface. The announcer may proclaim of Taris that "water is his element, like that of a fish" (ca. 2:01), but topside, Taris is visibly a human athlete, pushing heavy water with his powerful stroke.[151] A different view of Taris, however, is briefly intimated around 4:50,

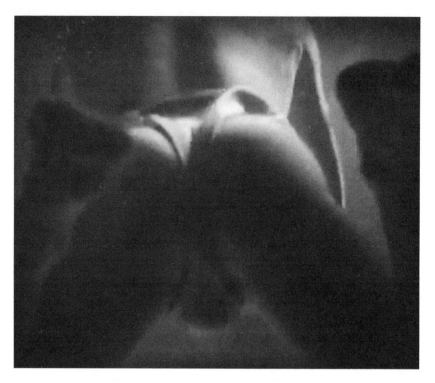

FIG. 1.17. Jean Vigo, *Taris, King of the Water*, ca. 7:36. When Taris turns in front of the camera, the underwater haze mutes the sequence's exhibition of his ass.

when the camera lets viewers glimpse his gliding body from below. The film lets us linger in the depths only after a didactic surface explanation of Taris's strokes, on the occasion when the narrator explains Taris's technique in his turns (ca. 7:15). At this point, the film abandons its didactic exposition to celebrate the beauty of the male body, just as the pool lets Taris shed the overcoat that returns him to everyday life at the end of the film. Similar to Man Ray's use of blur to hide female genitalia, the underwater atmosphere in *Taris* enables some muted exhibitionism, such as showing Taris's ass and hints of his genitalia in his bathing suit, when he executes a turn in front of the camera (fig. 1.17). In other shots, Taris melts into the atmosphere, outlined by a glow of light ornamented by clouds of bubbles, as in one shot of his powerful kick (fig. 1.18). In 1931, Vigo wrote to Jean Painlevé of the film, "Filmed above and below water. The creature and the environment are beautiful, but . . . Jean Vigo is not Jean Painlevé. And that's that!"[152] In contrast to Painlevé's exhibition of marine life to effect surrealist displacement, Vigo uses the underwater atmosphere to celebrate the human body.

FIG. 1.18. Jean Vigo, *Taris, King of the Water*, ca. 7:38. Taris's powerful kick illuminated and softened with underwater glamour.

"The creature and the environment are beautiful," as well in the underwater scene at an emotional climax in Vigo's *L'Atalante*. The subject of the film are two newlyweds, very much in love, yet struggling with the difficulties of married life. The husband is a barge captain named Jean played by Jean Dasté, and the young wife Juliette is played by Dita Parlo. The setup for an underwater scene occurs in the honeymoon phase of their relationship, when the couple are fooling around on the barge. Juliette tells Jean of a superstition that one sees the person one loves in water (ca. 16:28). Defying her superstition, the practical Jean sticks his head first in a bucket of water, then in the canal, and concludes, "I saw nothing." "You'll see something when you do that for real," she retorts, as she lovingly hands him a warm shirt (ca. 18:19).[153]

Juliette finds life as a barge captain's wife tiresome, with tense navigation shrouded in fog and early morning wakeups. Vigo uses the mists surrounding the barge that cause stress for the captain and his crew to steep Juliette in an atmosphere of longing, as in the image of her looking out into the fog (fig. 1.19). When the captain proves a reluctant and jealous escort to a dance, she runs

FIG. 1.19. Jean Vigo, *L'Atalante*, ca. 23:41. Juliette's longing expressed through a marine layer.

away. Separated, the couple miss each other desperately—she listens to sea shanties, and, at one point, the desperate Jean jumps overboard. Rather than suicide, however, Vigo shoots an underwater swimming scene where Jean searches in the depths. In this search, he sheds his tense topside demeanor, performing underwater the dance he had disdained on land. The haze even renders him elegant, and the camera brings us in close contact with his expressive countenance and his taut body. Eventually, as Juliette predicted, she comes floating toward him in a superimposed image, pivoting with a grace mixing the glide of water and the torque against gravity possible in air (see fig. 0.5).

As the scene concludes, Vigo returns to the expressive possibilities of the contrast between topside and underwater views he had used to transfigure Taris in his short. After Jean's underwater vision, the camera takes us topside and portrays a comic moment. Jules and the boy look for signs of Jean in the water, fearing he committed suicide, while Jean sneaks up and looks over their shoulder. Fittingly, the weather on the canal at this moment is beautiful, and Vigo uses a deep focus shot that could be an icon of linear perspective, as the side of the canal runs tranquilly in a straight line to a vanishing point on the horizon.

1.5. Adventures in Underwater Film until 1953:
The Constraints of the Cinematic Aquarium
and the Advent of the Underwater Eye

From Paton's *Twenty Thousand Leagues under the Sea* to the end of the 1940s, popular movies were made with dramatic underwater sequences of submarine adventure. Some were well-received by publics, but, among the films I have been able to locate, none received the acclaim of Paton's film, or Richard Fleischer's 1954 version of Verne's novel discussed in part 2. Until the postwar era, narrative films that did not, in contrast to surrealism, explore new artistic possibilities, largely kept to the format of the cinematic aquarium. While the first underwater views had been thrilling in 1916, by the 1930s and '40s, they had become commonplace. Narrative films used the dangers of salvage, rival divers, and sea creatures to engage their audiences with adventure in the depths, including oxygen shortage and the bends.

Helmet-diving adventures made on underwater film sets in the 1930s include *Sixteen Fathoms Deep* (1934), directed by Armand Schaefer, based on a short story about the community of Greek sponge fishermen in Tarpon Springs, Florida, by Eustace L. Adams, first published in the *American Magazine*. The first extended underwater sequence nods to the cinema of attractions, as a drunk tries to hide booze by drinking underwater. The cinema of attractions is evoked as well in the film's establishing sequence on the reef, which tracks the beautiful setting where fish flit for a mesmerizing thirty seconds, without an obtrusive overlay of sound. However, most of the spectators' interest derives from observing the dangers to the divers, in keeping with adventure narrative. In the terse words of *Variety*, "Plot chiefly delves with the tampering of the pumps and diving suits. Underwater stuff is well done and convincing."[154] The dangers are the work of a saboteur, Nick, played by Maurice Black. After two attacks on sponge divers below, the saboteur's air hose is fouled by a rock. Joe, the film's hero, played by Lon Chaney Jr., freedives and cuts him loose to save him.

Other helmet-diving adventures made on underwater film sets in the 1930s include *Below the Sea* (1933), directed by Albert Rogell, and his subsequent *No More Women* (1934). To salvage diving and sabotage, *Below the Sea* adds Fay Wray, playing Diana Templeton, a wealthy woman with an interest in marine science, who intrepidly dives in a helmet rig, where she gets the bends. She also uses a diving bell for underwater photography of sea creatures. In a climactic scene, a giant octopus breaks the chain tethering the bell to the surface,

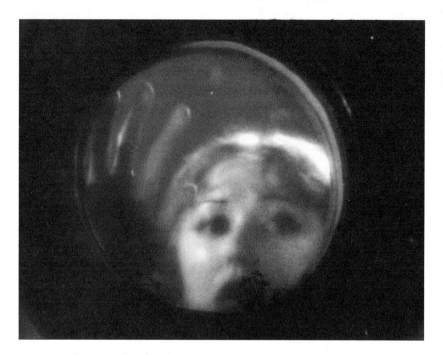

FIG. 1.20. Albert Rogell, *Below the Sea*, ca. 1:09:48. Fay Wray suffocating in a diving bell: hazy horror.

along with the bell's air hose, almost smothering her, before a diver / love interest played by Ralph Bellamy rescues her (fig. 1.20). Shot in black and white, the film also included four minutes in Technicolor of underwater life, but these minutes have been lost.[155] The 1935 *Mutiny Ahead*, directed by Thomas Atkins, featured "Diamonds, Dames, and Danger!," as was promised on a publicity poster. Four years after *Dracula*, Bela Lugosi played a jewel smuggler in Erle C. Kenton's *The Best Man Wins* (1935), again with adventures of salvage diving.

The predictable contexts where helmet diving was employed, along with the familiarity of the action shown below, limited these films' ability to hold spectators' interest in their underwater scenes. A good example of the problems ensuing from the limits of underwater exhibition before the portable camera and scuba, among other transformative innovations, is Ray Friedgen's *Killers of the Sea* (1937), a melodramatic documentary about Florida police chief Wallace Caswell Jr. As the "Foreword" of the title sequence introduces him, Caswell is an adventurer "of supreme courage" (ca. 1:40) on a "relentless crusade" in the Gulf of Mexico "against Killers of the Sea; which are rapidly destroying the game fish of Southern waters" (ca. 1:50). Underwater predators figure prominently in this film showing "man against Monster in his native

FIG. 1.21. Ray Friedgen, *Killers of the Sea*, ca. 36:53. The film depends on its voiceover to reveal that this blurry image depicts a "fairyland of eerie beauty."

element" (ca. 1:25). The camera person obtaining such views may have been American helmet diver, photographer, and best-selling author John D. Craig. Although Craig is not listed in the film, an underwater portion of the film did appear together with Craig's biography on the website Diving Almanac.[156] Craig had started to publish articles on his undersea experiences in advance of his full-length *Danger Is My Business* (1938), an immensely popular book that went through twelve English, two Dutch, and two German editions in 1938 alone.[157]

The first half of *Killers of the Sea* shows Caswell hunting topside. A frenetic melodramatic soundtrack accompanies his pursuit of a marlin, the landing of a hammerhead shark, and a lot of thrashing in whitewater, as he tries to subdue what the film portrays as a horrific marine animal: the bottlenose dolphin that is an enemy and competitor of the fisherman. When the film takes us underwater, blurry images of the depths are mismatched to the overwrought language of the voiceover narration. Thus, compare the "fairyland of eerie beauty" (ca. 36:53) to the ghostly, dim outlines actually shown in the film, where we merely see undulating seaweed (fig. 1.21). Or again, compare the dim outline of an octopus to the narrator's hyperbolic description of the creature as a "slimy, slithery hateful, a foul monstrosity in a grotesque world" (ca. 38:29).

Writing in *Billboard*, Sylvia Weiss remarked that the voiceover's "strained, mirthless description might be mistaken for a new form of torture," in a harsh review of this "stagy, uninspiring, artificial picture."[158] The most legible extended underwater sequence figures the octopus, where Caswell purportedly kills it, as the screen fills with ink (ca. 43:30), one of several underwater action sequences of killing in the film. The struggle to kill a tiger shark yields several impressive views of the shark's jaws from below, although the blurry action only fits together in the narrative created by the voiceover. The concluding struggle with the sawfish starts underwater. However, *Killers of the Sea* soon reverts to purple prose and topside thrashing until the sawfish slashes Caswell with its bill, spilling his blood, which could attract sharks, and hence putting an end to the drama.

In 1948, director Irving Allen remade *Sixteen Fathoms Deep* with the title *16 Fathoms Deep*, again featuring Lon Chaney Jr.—now in the role of villain rather than hero, however. This film shot in "Rainbow Springs, Fla and Marineland studios at St. Augustine, Fla," continued to use the format of the cinematic aquarium, but its underwater scenes were "photographed from a chartered submarine, and it was the "first feature length picture to be filmed in Ansco colour."[159] *16 Fathoms Deep* captivated audiences with its underwater views at a time when there were few underwater color photographs and marks the beginnings of innovations that, as we will see, would go on to renew the underwater film set across the next decade. The *Los Angeles Times* dismissed the plot as "old hat," but "[i]n spite of a trite story, action in the picture develops remarkably due to the environment. Underwater scenes showing the divers gathering sponges are unique in the present period."[160] Other contemporary reviews concurred, and the *Chicago Daily Tribune* suggested that the submarine realm threatened "at times to steal the show from the human participants."[161] The fascination of the underwater scenes in this film reached audiences around the world. The reviewer in the *Times of India*, for example, appreciated the "natural though sometimes frightening beauty" of the "oceanic water of the sponge-fishing village of Tarpon Springs, Florida."[162]

The film also innovated in expanding the types of action set below. Notably, the film shows extended sequences of sponge diving as a skilled form of fishing, well beyond the 1934 version. We learn about the sponge habitat, forks and sacks used to harvest them, and how they are cleaned and sold. We also learn technical details about helmet diving, from the gear to such basic aspects as moving through the weighty atmosphere of water to professional tricks, such as how a diver floats himself to the ladder once he surfaces by letting his suit get inflated with air, a maneuver that requires some skill so as not to over-fill the suit. These details are explained by the character Ray Douglas, an "ex-Navy diver who hits the beach at Tarpon Springs in try to get on a sponge boat," played by a young Lloyd Bridges.[163] The *Daily Boston Globe* found the voiceover narration "at times" "forced and not too agreeable" but nevertheless reviewed the film as "unusual, thrilling and pictorially distinguished."[164] Bridges's understated voiceover became a distinctive feature of *Sea Hunt* (1958–61), the series that catapulted him to a star known worldwide for his underwater exploits.

Beyond showing diving techniques, the film used the underwater film set to create suspense, with situations that would be amplified throughout the scuba era. In the sequence with young men diving for a Greek cross, both underwater obscurity and the veil of the surface briefly mask the identity of the diver who wins the competition. The attack of a shark on a helmet diver creates suspense in a long tradition of confronting divers with ferocious preda-tors. The sight of a shark snapping at the helmet of a diver gives the scene ur-gency that enables us to overlook the lack of continuity between the studio shot of the hero in a round helmet and the helmet used by the stunt diver in the flow of the scene. In the final sabotage, suspense is created through a threat to the diver's air supply, contrived by the villain who has stealthily cut the "klugers" thrown over the propeller to protect the diver's air hose from it, so that it is only a matter of time until the hose is entangled in the propeller clip. Navy diver Douglas, with the keen sixth sense that would become the hallmark of Mike Nelson's underwater craft, gets a bad feeling and surfaces, in contrast to Athos, a veteran old sponge diver. When Athos's bubbles stop rising, his son dives to save him, but a giant clam catches the son's leg, and the son tragically drowns before his father's eyes.

In the early 1950s, portable underwater cameras together with scuba started to enter the entertainment film industry in the United States. Cousteau had made his first short documentary with scuba in 1943 with a portable underwa-ter camera, which he helped invent, and both he and Hass would revolutionize the styles of underwater filming, as I discuss in part 2. In 1951, Cousteau's scuba technology entered the United States, and in the same year, Chuck Blakeslee and Jim Auxier launched the *Skin Diver*, the first magazine of recreational div-ing. That same year, 20th Century Fox released a film credited with bringing the portable underwater camera to entertainment film: Lloyd Bacon's *The Frogmen* (1951) about underwater demolition teams (UDTs) during World War II. According to the *New York Herald Tribune*, "For the first time they

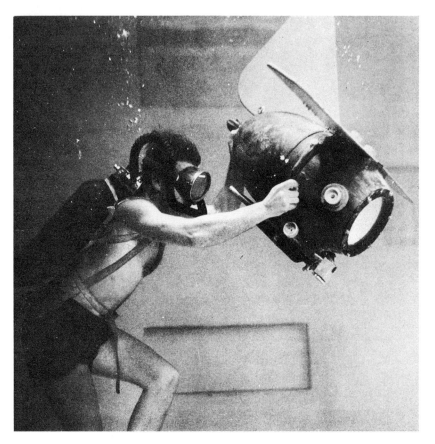

FIG. 1.22. "Fishy Camera," the Aquaflex, photographer unknown. Titled thus in C. B. Colby, *Frogmen: Training, Equipment, and Operations of Our Navy's Undersea Fighters* (New York: Coward-McCann, 1954), 16.

[studio technicians] used a revolutionary underwater motion picture camera, independent of both air supply and electric cables linking it with the surface. The camera, known as the Aquaflex, is fitted with detachable wings and vertical rudder, weighs 107 pounds, and carries its own air supply."[165] The Aquaflex had been invented in the late 1940s by adapting the portable French camera of the 1940s, the Caméflex, with "a blimp for sound shooting and an underwater shell." The Aquaflex was "the first camera specially designed for submarine shooting used in Hollywood and by the French and American navies" (fig. 1.22).[166]

The Frogmen received positive reviews when it appeared, as all newspapers agreed, in the words of the *New York Times* headline, that it offered a fascinating glimpse into the work of the "Unsung Heroes of Naval Action."[167] It

FIG. 1.23. Lloyd Bacon, *The Frogmen*, ca. 1:30:33. Suspenseful action gripped viewers in this climatic combat filmed using the Aquaflex.

featured extensive underwater sequences: first of the UDTs swimming with masks and flippers, which were not commonly used at the time, and then with scuba. These sequences were filmed "on location at the Amphibious Base at Little Creek, VA., Key West, Fla., and off the Virgin Islands."[168] Or rather, in the words of the review in the *Christian Science Monitor*, it might be "more proper to say they spent seven weeks 'in' location for they filmed more than 90 percent of the picture under water."[169] Perhaps the article was exaggerating, and it was written before the film was released in the summer of 1951, but in any event, underwater sequences still occupy a remarkably large portion of the screen time.

The takeaway from reviews is that audiences appreciated both information on the under-known activities of frogmen and the film's extensive, detailed, and diverse portrayal of human action underwater. "Purely from a technical standpoint, the underwater photography is truly remarkable," commented the *New York Times*. "Fantastic as it may be, there is tremendous excitement when the Americans fight a subsurface battle with Nipponese frogmen."[170] In the screengrab from this sequence, we see a Japanese diver, using a rebreather of the type used in World War II, locked in combat with one of the frogmen (fig. 1.23).

The action is enthralling on its own, and the only soundtrack in this extended final combat is of bubbles and the divers breathing. Nonetheless, while the action is dramatic, the imagery largely remains within the landscape format, using the focus of a medium shot, and does not explore more radical possibilities for framing and angles in the three dimensions of viewing available.

Underwater cinematography, in contrast, would become more fluid and offer multiple perspectives that brought the environment as well as the action to life in Robert Webb's *Beneath the 12-Mile Reef* (1953). This film, about competition between white redneck and Greek immigrant sponge divers, was shot "entirely on location at Tarpon Springs and Key West, Florida" with the Aquaflex.[171] *Beneath the 12-Mile Reef* is the third film to be shot in the wide screen format of Cinemascope, and it was made in Technicolor. In contrast to *The Frogmen*, *Beneath the 12-Mile Reef* features a credit for an underwater cameraman—Till Gabbani, who may have used scuba, although I have not found this detail confirmed. Certainly, Gabbani would use scuba when he took the job as chief cameraman for Disney's production of *20,000 Leagues under the Sea* discussed in part 2, released in 1954.

The subject matter of Greek-American sponge divers was not new, and the *New York Times* film critic Bosley Crowther said that "the drama developed in the screen play is hackneyed and banal."[172] Crowther's reviews were widely read, and, as will become evident as I quote from them in illustrating reception of films from the 1940s to the 1960s, his tone was tart and pithy. His view of the plot, as well as his general reluctance to be effusive makes his words on the exceptional underwater sequence in *Beneath the 12-Mile Reef* all the more noteworthy:

> [The] anamorphic lens goes under water to picture marine life around the tropical Florida keys and comes up with scenes of the floor of the ocean that are bigger than any of the sort you've ever seen. Coral reefs and jewel-colored fishes and gentlemen in bubbling diving suits loom through the opalescent water that stretches from wall to wall across the front of the theatre within the frame of the panel screen, and the inevitable octopus slithers out of the sub-aqueous gloom. Fishermen rake rubbery sponges off the ocean bed, and everything looks both large and fearful down there beneath the watery blue-green sea.[173]

Although Crowther does not include the portable camera in his assessment, its ability to portray multiple points of view underwater, along with the film's color, shape his enthusiasm. *Variety* was more attentive to the "underwater photography," naming Till Gabbani, and finding "punch in the display of underwater wonders," as well as "a strong air of expectancy in the water footage that sustains a feel of action and creates suspense."[174] *Beneath the 12-Mile Reef* was nominated for the Grand Prize of the Cannes Film Festival, and cinematographer Edward Cronjager was nominated for an Oscar for the best color cinematography.

As discussed in part 2, the underwater exhibition enabled by scuba and the mobility of the portable camera would dazzle in *The Silent World*, which won the Palme d'Or at Cannes and received an Oscar for best documentary. With such enabling technologies, as well as the vast underwater expertise of Cousteau and the savvy eye of Louis Malle, *The Silent World* amplified shot possibilities and underwater locations well beyond the exhibition of a single setting in *Beneath the 12-Mile Reef*. Yet Webb's film and Gabbani's camerawork nonetheless give spectators a more varied and detailed sense of the environment than ever before and expand the types of action that can be shown, even including underwater views in a romantic swimming scene, along with helmet-diving adventure.

Particularly notable among the amplified shot possibilities are sequences conveying the diver's point of view. Two scenes at the apex of the drama illustrate this perspective: when the hero's father, Mike Petrakis, played topside by Gilbert Roland, first visits the twelve-mile reef in his fatal dive, and then when Tony steps into his father's role and explores this dangerous environment. In both scenes, the camera not only exhibits the reef using a landscape format in medium shots but also offers views from a range of perspectives and scales. Thus, for example (ca. 44:00), we watch Petrakis move carefully around the reef, disappear behind some coral, and then draw near again. Particularly dramatic is a close-up looking down on his helmet as he climbs up a big coral formation (ca. 45:57), just before he falls off the edge.

After his father's death, Tony, played topside by Robert Wagner, takes over his father's work below. His initiatory dive creates narrative suspense given the dangers of the reef, which Tony had heard his father describe but has never before seen (ca. 1:27:40). The film further adapts Tony's point of view discovering the reef together with the spectators. It uses this perspective in shot/countershot pairing, matching views of Tony gazing with what he sees. The shot/countershot recurs as Tony finds and harvests sponges; indeed, the movements of the camera even blur the vistas at times, as if suggesting Tony's scanning, exploratory gaze. As he looks down to start to search for sponges,

FIG. 1.24. Robert Webb, *Beneath the 12-Mile Reef*, ca. 1:30:01. A portable camera brings to view-ers different distances and orientations, opening up a realm of expressive possibilities for shots below. In this screengrab, we are looking in the direction of Tony's gaze: toward the seafloor.

FIG. 1.25. Robert Webb, *Beneath the 12-Mile Reef*, ca. 1:30:08. And in this screengrab, we see the object of the gaze: sponges.

for example, the film shows the downward direction of his helmet and then the seafloor he is examining (figs. 1.24, 1.25).

In its ability to descend with divers and show their mobile viewpoint, *Be-neath the 12-Mile Reef* offers spectators a new relation to the submarine envi-ronment. Along with witnessing the reef as a fascinating spectacle that passes before our eyes, we shadow the characters' agency immersed in the depths. The portable camera enables a mobile gaze around the underwater environment, and film audiences join characters in looking with *an underwater eye*.

1.6. Submarine Materiality and the Pleasures of Artifice: Cecil B. DeMille's Silken Kelp and Mechanical Squid

In my survey of the first era of underwater film, I have described two different ways filmmakers use submarine materiality, whether in documentary or fantasy: to expand our glimpse into unknown areas of the planet and to open new domains for fantasy, whether in popular spectacles such as Kellerman's or the more rebarbative explorations of *piscinéma* by Man Ray. In conclusion, I give one more example of their interest in the underwater film set as the location for creating beguiling artifice that has little connection to the reality of underwater environments. I end with this emphasis to remind readers that submarine materiality remains a potent resource for fantasy, even if a great deal of part 2 focuses on works celebrating new dive and camera technologies as expanding access to the underwater environment.

Take, for example, the underwater conflict that culminates Cecil B. DeMille's Technicolor *Reap the Wild Wind* (1942). DeMille made this swashbuckling antebellum saga of shipwreck and treasure hunting set in Florida to take advantage of the success of *Gone with the Wind* (1939). The trailer boasted its "vivid realism," and grand-scale romance. Realism is in the eye of the beholder, and when it comes to underwater scenes, the realism was to recreate a recognizable set beneath the waves rather than to create a facsimile of an undersea environment.

The primary underwater film set for *Reap the Wild Wind* was "what is known as the Big Tank at Paramount Studios," according to actor Ray Milland. Milland played the upright male lead, attorney Steven Tolliver, who duels with the more shady Captain Jack Stuart, played by John Wayne, for the affections of Loxi, the spunky female ship salvager played by Paulette Goddard. Milland recalls, "Down there they had built a marine wonderland: the hull of a wrecked ship, strange and jagged rocks, a slowly moving aqueous forest. And caves, dark and frightening."[175] The Turner Classic Movie website suggests the film used underwater footage captured in diverse locations. It was reported "that underwater sequences were filmed at the Pacific Marine Museum in Malibu, CA, although modern sources suggest that the scene with the giant squid was filmed in a tank at the Paramount studios. According to an article in *New York*

Herald Tribune, preparation for the underwater scenes took two months, and an additional two months were spent in shooting the underwater *Southern Cross* scenes. DeMille also used the . . . water tank at the United Artists studio, which had to be enlarged to shoot the ship sequences, and the 'Old Hawks' Tank' and 'The Water Way' at the Columbia Ranch in Burbank, CA."[176]

The entire film, topside as below, reveled in fantasy rather than creating realistic settings. As Crowther wrote, "*Reap the Wild Wind* bulges with backgrounds which have the texture of museum displays. Rooms reek of quality and substance, gardens look like the annual flower show and the scenes of ships on the high seas exude a definite suggestion of salt air."[177] Below, the museological quality turns to artistry, as DeMille created kelp using "yards and yards of iridescent silk that would weave about in the ocean depths. Unfortunately, the color of silk changes in sea water so DeMille's production team, headed by cinematographer Victor Milner, had to use fresh silk every day, then dry it out, and have it dyed back to its original shade for reuse."[178]

This same artifice characterizes the hold of the wreck where the film's dénouement is set. It is spacious, with fabric also appearing as part of the cargo and a startling variety of fish (not necessarily native to Florida, such as the Pacific orange garibaldi). "According to an article in *New York Herald Tribune,* fish and lobster were culled from the Pacific Ocean in Malibu with the permission of the California Fish and Game Commission, and were used to stock the tanks."[179] This confined space scaled to the human body reduced the diffuse, disorienting quality of underwater optics—a practice anticipating the Disney studio's wreck set in *20,000 Leagues under the Sea,* discussed in part 2.

The wreck set provides the backdrop for deciding the contest between Tolliver and Stuart, when they dive together to gather information about whether a stowaway young woman, the heroine Loxi's cousin, has died there, a victim of Stuart's machinations to sink the *Southern Cross.* In this picture "splashed . . . with every color," to quote Crowther, DeMille uses the atmosphere to emotional effect, as it washes out the formerly vivid colors of the fabric once contained neatly in bales that now streams chaotically in the wreck. Their pallor anticipates confirmation that Loxi's cousin has drowned, through identification of her shawl, whose pale yellow fringes undulate like long blond hair, while fish, like ghosts, glide by. The expressiveness of color in water extends as well to the giant squid that attacks both Tolliver and Stuart at the climax of the wreck scene, where, in Crowther's words, the men can "manifest the stuff of which they are made." Although the scene was filmed before Hollywood

FIG. 1.26. Cecil B. DeMille, *Reap the Wild Wind*, ca. 1:53:04. Artifice delights viewers in this scene of combat with an obviously fabricated giant squid.

had access to the portable camera, it holds drama, owing to its fantastical squid with its invasive tentacles as well as to the magnificent set steeped in an atmosphere of "greenish opalescence" (fig. 1.26).[180]

According to the Turner Movie Classics website, DeMille initially planned on using a living creature. "But he ended up settling for a huge mechanical squid, which was operated by a twenty-four button electrical keyboard with a complex system of hydraulic pistons used to activate the thirty-foot tentacles of the squid."[181] The squid's color is exuberantly artificial, for it is an angry orange-red, like a cooked lobster, with a green, cold giant eye. Description of filming makes the work sound comic, but it must have been arduous and chaotic as "DeMille communicated by a loudspeaker to direct the technicians in the movement of the creature while at the same time barking orders to Milland and John Wayne underwater through telephone wires in their helmets (Wayne would later battle another underwater sea creature in *Wake of the Red Witch*, 1948)."[182] DeMille was so concerned with the outcome of the scene that he donned "a diving suit and personally attended to the staging of Milland and Wayne's subsea fight with the giant squid."[183]

Although it had nothing like *Gone with the Wind*'s success, *Reap the Wild Wind* was included in the "All-Time Top Film Grossers" [Over $4,000,000 US-Canada Rentals]" in a *Variety* list from January 8, 1964, which *Gone with the Wind* topped, and which included *20,000 Leagues under the Sea* as well.[184] We can only presume that audiences appreciated the film's extravagant fancy, including in its underwater scenes. According to Turner Classic Movies website, the film when it appeared was "hailed by reviewers for its technical achievements and won an Academy Award in the category of Special Effects (Farciot Edouart, Gordon Jennings, William L. Pereira and Louis Mesenkop). It was nominated for Academy Awards in the following categories: Cinematography (Color), Victor Milner and William V. Skall; and Art Direction/ Interior Decoration (Color), Hans Dreier and Roland Anderson, George Sawly."[185]

The most evident impact of scuba and the portable cameras on moving imagery during the 1950s was vastly to expand access and enable film to exhibit new areas of the sea. At the same time, filmmaker's interest in submarine materiality as a resource for fantasy effects continues into the revolutionary era discussed in part 2, whether it was the science fiction settings of Verne in Disney's *20,000 Leagues*, which feature another mechanical cephalopod; horror films such as *Creature from the Black Lagoon* (1954), directed by Jack Arnold; or George Sidney's underwater dance fantasia, *Jupiter's Darling* (1955), starring Esther Williams. This fantastic lineage will flower in the liquid fantasies featured in part 3, where submarine materiality and indeed *piscinéma* offer potent resources for creativity in general audience narrative film.

2

The Wet Camera, 1951–1961

2.1. The Post–World War II Revolution
in Cinematic Access

In 1956, the Cannes jury awarded the festival's top award to *The Silent World*, codirected by Jacques-Yves Cousteau and Louis Malle. From the perspective of cinema history, Cousteau and Malle's film is an outlier in a prize list that reads as a roster of illustrious directors who shaped the language of film. In 1956, *The Silent World* won over submissions by Alfred Hitchcock (*The Man Who Knew Too Much*), Henri-Georges Clouzot (*The Picasso Mystery*), Ingmar Bergman (*Smiles of a Summer Night*), Satyajit Ray (*Pather Panchali*), and Akira Kurosawa (*I Live in Fear*).[1] Why did the jury put this undersea travelogue, the first and one of the only two documentaries ever to receive the Palme d'Or prize, at the pinnacle of contemporary acclaim?[2]

The usually trenchant French film critic André Bazin opened his review declaring that to undertake criticism of *The Silent World* was "ridiculous [*dérisoire*]," as "the beauties of the film are first of all those of nature and one might as well undertake criticism of God."[3] What Bazin downplayed—and perhaps did not know, given the novelty at the time of Cousteau and Malle's techniques for filming underwater—was how the film's image of nature was created through these directors' artistic deployment of new innovations in diving and underwater film.

The technologies enabling such creativity date to World War II and the immediate postwar era.[4] They include, notably, scuba and the portable camera. Modern scuba was vastly less dangerous and expensive than the type of CCOUBA (Closed-Circuit O_2 Underwater Breathing Apparatus) that Williamson procured from the British military for the divers in the silent *20,000*

Leagues under the Sea (1916).[5] Once invented, scuba offered new access for science and exploration. Revolutionary technologies specific to film included water-housing, lenses, and flash compensating variously for the distortions of underwater optics, low-light conditions, and the increased pressure of depth. The first wet suits offered warmer and more flexible insulation than helmet diving suits, and swim fins, underwater scooters, and sleds allowed people and equipment to move rapidly through water's dense environment. Further expanding the reach of underwater photography and film were more general developments in moving image media, such as wide-screen processes and color film that could be used in all types of camera. In addition, the burgeoning medium of television enabled viewers at home to enjoy spectacles of the most inaccessible environment on our planet.

These new technologies created new access to a vastly expanded range of marine ecosystems, in different waters and at different depths, as well as the ability to portray them with greater clarity and with new kinds of movements, new camera angles, and in mesmerizing color. Dive technologies, moreover, were developed by the same people who pioneered underwater image-making, and the hunt for vivid images of the undersea prompted reciprocal innovation. American dive pioneers Zale Parry and Albert Tillman indeed suggest that "[u]nderwater photography was certainly the major driving force that brought scuba and the Aqua Lung into existence."[6]

A technological history of the revolution in dive and underwater cameras is outside the scope of this book, but it is worth noting just how international this moment was in the history of underwater access. Engineers and diving teams from nations in the vanguard of industrial modernity, notably with coastlines on warm water, were experimenting with the same problems, competing with each other, and devising related, sometimes slightly different solutions.[7] Cousteau, for example, although he does not mention it, was competing with the Austrian Hans Hass, less well known today in the Anglophone world.[8] Hass, too, was a scuba pioneer: he invented a rebreather that used chemicals to scrub carbon dioxide from the diver's exhaled air rather than relying on the tanks of compressed air that Cousteau used.[9] Hass's rebreather was taken up by professionals, as it enabled diving at deeper depths than did compressed air, although it was more dangerous and required greater skill. Hass also experimented with underwater technologies for photography and film. He developed the Rolleimarin housing for the Rolleiflex, which became the gold standard for professional photographers after it was patented in 1954.[10] As the careers of Cousteau and Hass illustrate, people worked together in international teams.[11] Cousteau collaborated with the legendary MIT electrical engineering professor Harold Edgerton in developing underwater lighting, working with the Woods Hole Oceanographic Institution in Massachusetts and subsequently the Scripps Institution of Oceanography in La Jolla, California. Hans Hass teamed up with the BBC to develop a six-part TV series, *Diving to Adventure*, screened first in April–May 1956, which appears to have been the BBC's first miniseries about the ocean, half a century before *The Blue Planet*.

The impact of these developments on the practice and imagination of the submarine realm was comparable to the impact of the aquarium when it emerged. Indeed, we can see numerous parallels between the impact of the aquarium and of new imagery captured beneath the surface of the water in the post–World War II years. Like the aquariums a century earlier, these new exhibition technologies affected both leisure and professional communities, as well as creating intersections between them. The aquarium influenced the development of modern oceanography and marine biology; it also stimulated amateur coastal naturalism and household cultivation of marine tanks. Similarly, underwater photography and film aided the expansion of science beneath the seas and promoted aquatic recreation, from diving and snorkeling to fishing, and surfing.[12] Amateurs also took up underwater photography, often in coordination with other aquatic activities.

The role of diverse media in popularizing new marine knowledge is another parallel with the aquarium era. Gosse's *The Aquarium: An Unveiling of the Wonders of the Deep Sea* (1854) was an early example of nineteenth-century books about seaside naturalism, by the man who had also invented the aquarium as public exhibition form. To draw a parallel with Cousteau, in 1953, three years before the film *The Silent World*, Cousteau, with Frédéric Dumas, wrote an acclaimed illustrated book with the same title, published both in France and in English translation.[13] The book, which recounted Cousteau's experience diving and filming, contained the first color photographs showing the presence of red light in the depths, a revelation only possible with artificial underwater lighting, which Cousteau helped develop. Periodicals also brought underwater photography, film, and diving to public awareness, stimulating amateur imitation. Cousteau published several articles in *National Geographic*, which helped sponsor his research: the magazine had a wide reach, with a total international circulation of over 1.8 million in 1950 that had grown to over 2.3 million by 1959.[14] Cousteau also toured, giving slide lectures, which were televised by the BBC as early as 1950.[15]

Historian Helen M. Rozwadowski has told the story of how science and the imagination worked together in both nineteenth- and twentieth-century revolutionary eras of underwater access to create intense psychological bonds for general publics with the ocean's inaccessible spaces. Rozwadowski includes leisure activities like diving and recreational photography, along with science, as critical to such expansion, noting that fostering the imagination may be even more important than scientific data in leading people to conceive of realms so far beyond empirical experience.[16] She also stresses the importance of entertainment in delivering "knowledge of the oceans, no less than did work in, on, and under the ocean."[17]

Films set underwater, both documentary and fictional works, proliferated, and, by the end of the 1950s, TV started to feature shows with underwater subjects as well. Some of the first full-length documentary works include Hans Hass's black-and-white *Under the Red Sea* (*Abenteuer im Roten Meer*) (1951) and his color film *Under the Caribbean* (*Unternehmen Xarifa*) (1954), as well as Folco Quilici's *Sixth Continent* (*Sesto continente*) (1954) and Cousteau and Malle's *The Silent World*. Fictional films in the same years range from the previously discussed Robert Webb's *Beneath the 12-Mile Reef* (1953) and Lee Robinson's *King of the Coral Sea* (1954) to Roger Corman's *Attack of the Crab Monsters* (1957) and Andrew Marton's *Underwater Warrior* (1958) as well as TV series such as *Sea Hunt* (1958–61), produced by Ivan Tors. In the period 1953–61, I have identified forty-four narrative films and two TV dramas with significant underwater settings.[18] These settings were primarily undersea, although some were freshwater, most famously the Amazonian lagoon of *Creature from the Black Lagoon*.

Within this corpus, I have chosen to discuss influential works that expand the imaginative possibilities of the underwater film set. These works fit the category of spectacle at the time they appeared, seen by audiences of millions across the globe. *20,000 Leagues under the Sea* (1954) was the Walt Disney studio's first live-action film. Directed by Richard Fleischer, it shows how filmmakers constructed a new vision of the underwater environment as a realistic setting for complex human drama, using submarine materiality for mood but, like *Reap the Wild Wind*, importing stage conventions into the depths. *The Silent World* (1956) expanded types of environments used as settings, imbuing different zones of the undersea with a gamut of emotions and moods. The glamorous new sport of diving, including a female underwater adventurer, captivated audiences in the BBC TV miniseries *Diving to Adventure* (1956), which starred Hans and his wife Charlotte Hass, née Baierl, known by her nickname Lotte in her public dive and film career. Reusing footage from Hass's films, the series also showed audiences the marvels of the undersea, from dazzling coral reefs to sharks. The first fictional TV series with underwater scenes, *Sea Hunt*, devised narrative conventions for presenting action-packed adventures underwater that yielded 155 episodes over four seasons. In these adventures, *Sea Hunt* introduced audiences to the diverse uses of the undersea in contemporary commercial, military, and leisure activities, communicating technical knowledge with its innovative compressed poetics of suspense. Also important was the main character, Mike Nelson, an ex-navy frogman diver hero, taking under the waves what I have elsewhere called the craft of the mariner. In the words of dive historian Eric Hanauer, "The exploits of Mike Nelson (played by Lloyd Bridges) make the character a role model for a generation of wanna-be and will-be scuba divers."[19]

The newly accessible underwater realm was so captivatingly different that it created for some viewers an impression that it was at once poetry and fact. In *The Undersea Adventure* (*L'aventure sous-marine*) (French 1951, English 1953), the most philosophical book in the undersea literature of the immediate postwar era, Cousteau's co-diver and cowriter Philippe Diolé suggested that the reality of the underwater environment created new domains of fantasy for the imagination. Diolé devoted an entire chapter to the poetry of the depths, where he remarked that film was the ideal medium for their exhibition and that "[s]ubmarine photography has had, from the very beginning, a poetic value." Diolé gave the documentary imagery in Cousteau's black-and-white shorts from the 1940s an emotional tinge: "Films like those of Cousteau and Tailliez are not only faithful records of the marine world, but images full of what one might call the residue of dreams. They can convey to an audience completely ignorant of sea life, its whole range of poetic feeling."[20] Bazin expressed a similar sense when he saw *The Silent World*, describing what he called "the aesthetics of undersea exploration, or if one prefers its poetry," as "an integral part of the event."[21]

As such comments suggest, the line between documentary and imaginative representations in works of this revolutionary era blurred easily, for different types of reasons. In discussing his contribution to *The Silent World*, Malle recalls that he came to the film with youthful zeal about documentary authenticity. Inspired by the "austere" "view of film-making" of director Robert Bresson "to stay absolutely pure," Malle faulted Cousteau for "resorting to the techniques of what would be called today docudrama."[22] While respecting Malle's vision, we can grasp why, beyond the quest for popularity, Cousteau used techniques Malle denigrated as "conventional spectacle." Had Malle and Cousteau been filming an iconic, or even accessible, environment, such as Paris or the American West, they could have assumed audience familiarity, at least with images of such places, and hence had more freedom to leave familiar conventions behind. But instead, Cousteau was looking for ways to settle audiences in showing them the limits of diving with compressed air or the actual hull of a wrecked ship on the seafloor.

Throughout the spectacles discussed, image makers use extremely diverse reference points to make audiences at home in the new seascapes they have the ability to exhibit. These reference points range from pre-Romantic and Romantic aesthetics of landscape, which Hass, like Cousteau and Malle, steeped in water, and age-old fantasies of beautiful, bizarre, or fearsome undersea life to avant-garde poetics that were in some way analogous to the disorienting (from a terrestrial standpoint) physical conditions and creatures undersea.[23]

Despite documentary's use of dramatic techniques, the films of both Cousteau and Hass respected the integrity of the environment being filmed. "All shots shown were genuinely filmed in the Red Sea," is the first line of the first title of Hass's 1951 film *Under the Red Sea* (ca. 1:29).[24] *Sea Hunt*, in contrast, had no qualms about weaving together as one environment submarine scenes shot in different locations, or, alternatively, using a single location for scenes that the script set around the world. These places included the tanks of Marineland of the Pacific, as well as the waters off Southern California, Florida, and the Bahamas.[25] Further, *Sea Hunt* was famous for some implausible features, notably Nelson's solo heroism, violating the buddy system that formed the foundation of scuba protocol. Documentaries, in contrast, were careful to follow dive practice. We will see that *The Silent World* emphasized the importance of the buddy system and the danger when it is not observed in the episode when André Laban almost dies from nitrogen narcosis. After the intoxicated Laban pulls out his regulator, his dive buddy Falco puts it back in but then makes the mistake of sending the disoriented Laban to the surface alone. In this solo ascent, Laban forgets to make a decompression stop, and surfaces with decompression sickness.

Presumably, we are watching a simulation, because if Laban were being filmed when he made this mistake, the cameraman would have stopped him at the level where he needed to decompress, with safety taking precedence over drama. Cousteau and Hass respect the foundational documentary ethos of showing what could happen in the underwater environment; however, the events could be simulated or edited to transform them from their original capture. In such constructed representations of actual danger, spectators recognized zones where the line between fact and fiction was tenuous. Commenting on reactions to portrayals of near-death experiences in Hass's *Diving to Adventure*, the *Daily Mail* television reviewer Peter Black wrote, "Some of my critical colleagues . . . are becoming rather hoity-toity about the Hass series. They point out that when Hass says: 'Now this is where I am nearly drowning,' he is not sticking to the facts, since he could have been rescued at any time by the cameraman who was photographing his struggles."[26]

The leading French film theorist of the era, André Bazin, teased out another delicate balance in documentaries of extreme environments applicable to the films of Hass, Cousteau, and Malle. Such films at once showed unprecedented, exceptionally risky undertakings and captured images of these expeditions representing their adventure. Bazin pondered this balance in his review of Thor Heyerdahl's film *Kon-Tiki* (1950), which he characterized as "the most beautiful film, except it does not exist!"[27] Praising Heyerdahl's lively popular book, *The Kon-Tiki Expedition: By Raft across the South Seas* (1948), Bazin remarked that given media hierarchies for representing reality in 1952, "the only account worthy of the undertaking was on film." However, the expedition's amateur filmmaking did not capture suitably powerful imagery. In Bazin's words, "blurry and shaky images" appealed to viewers as the "photography . . . of danger," rather than documenting what was observed on the expedition— in the case of the event referred to here, an encounter with a shark. That Bazin did not question the accomplished, professional imagery of *The Silent World* indicates the film's success in striking the balance for audiences of the time between the truth of representation and an aesthetic control exhibiting events and the environment with clarity and force.

A factor blurring the boundary between documentary and fictional film with implications for both genres was the absence of what Malle called "rules," in his previously cited statement, "[w]e had to invent the rules—there were no references; it was too new."[28] Given that this environment hitherto had not been seen, let alone filmed, image makers were experimenting with creating a new repertoire of conventions. When they invented an appealing or gripping convention, other underwater filmmakers picked it up, to the extent that it did not violate basic tenets of their genre—no science fiction creature like Gill-man appeared in *The Silent World*. The character of the adventurous diving beauty would enter underwater narrative film through documentary. The model for this role was Lotte Hass, who starred in Hass's films from the early

1950s. This character appeared in narrative cinema as early as the Italian Francesco De Robertis's *Mizar* (*Frogwoman*) (1954) and the Jane Russell vehicle *Underwater!* (1955).

Successful cinematic innovations further migrated across the boundaries from film to TV. Indeed, episodes in the influential Hass series *Diving to Adventure* were in large measure made up of travel and dive footage from his previously released films. At the same time, the constraints of the television medium's small-screen, black-and-white format shaped the footage selected. The Hass TV shows, for example, entertained with dramatic animals that filled the entire screen, such as sharks and rays. They also took up an aspect of Hass's films well suited to TV, which was his portrayal of diving as an easy, glamorous lifestyle that was part of carefree, youthful beach culture. By emphasizing such human interest, *Diving to Adventure*'s episodes could compensate for the small size of television's visual display. Further, these nonchalant young divers made themselves approachable, contrasting with the distance created by the professional austerity of Cousteau and his fellow navy team members.

Both the fascination of the beach lifestyle and the work of the professional diver would be developed in *Sea Hunt* as well as in subsequent underwater films. One innovation of *Sea Hunt* in response to the demands of a weekly TV series that would then pass to the medium of film was its pared-down adventure formula. *Sea Hunt* enthralled audiences with few props in a scant twenty-five minutes, using basic physical dangers of the underwater atmosphere, such as running out of oxygen or decompression sickness. Such taut suspense created from elemental danger, even as it was suited to TV, would be picked up and used in underwater adventure films as well.

While viewers luxuriated in the underwater scenes, film companies and newspapers promoted films by emphasizing the difficulties of underwater moviemaking, giving a vivid sense of unexpected challenges.[29] Articles by Cousteau describing such difficulties appeared in *National Geographic* in the early 1950s, and the *New York Times* also published an article on the subject after *The Silent World* (1956) premiered in the United States. Months before the release of *20,000 Leagues* (December 1954), *Life* published an article with vivid illustrations: "A Weird New Film World: Jules Verne Movie Produces 20,000 Headaches under the Sea" (fig. 2.1). The *New York Times* featured an article on the same subject in March 1954. At the time of the release of *20,000 Leagues*, Disney aired a one-hour TV program called *Operation Undersea* (December 8, 1954), where submarine facts were integrated with details about making *20,000 Leagues*. A version of the *Life* article with different illustrations

FIG. 2.1. Peter Stackpole, "Underwater Filming of Walt Disney's Production of *20,000 Leagues under the Sea*," photograph. The LIFE Picture Collection via Getty Images. Photo used in "A Weird New Film World: Jules Verne Movie Produces 20,000 Headaches under the Sea," *Life*, February 22, 1954.

appeared in the British film fan magazine *The Picturegoer* ("20,000 Headaches," July 30, 1955), after the release of the film in the United Kingdom in spring 1955.

Difficulties mentioned in these articles ranged from the weather to the technology to the challenges for large film crews to ensure safety and quality production. To cite the introduction to the *Life* article, "In the Caribbean off Nassau, Bahamas, 83 actor-divers, cameramen, grips, propmen, professional salvage men, lifeguards and directors are trying to make their CinemaScope version of Jules Verne's science-fiction classic. . . . Daily the crew has to cope with choppy surface, undersea turbulence, leaky air valves on diving equipment, a complicated communications system, and sunlight that seems invariably to hide whenever all the other conditions are right."[30] As Fleischer described in a 2003 featurette, *The Making of 20,000 Leagues*, divers had enough air to stay under for one hour, which was measured from when "the first man went on air while he was still on the barge that we worked off of out in the water until the last man came out of the water." Along with the crew, there was one safety man for every two divers and crew members, and each star had a safety man for himself. "By the time we got to the location," Fleischer recounted, "we had just a few minutes to shoot the scene."[31]

Visual exhibition working together with narrative drama were key components to the new submarine fantasies created by moving images during the revolutionary years covered in this part of the book. In Franziska Torma's words, "The act of seeing, as opposed to the act of listening, is crucial to the history of underwater film."[32] The primacy of visual representation showed the authority of photographic imagery, both still and moving, as the medium trusted above all in this timeframe to document reality (in contrast to the primacy of writing for European cultures in the global age of sail). However, these works did not rely on visual exhibition alone. Soundscapes too cued audiences as to mood. Hans Hass went so far as to place speakers on the seafloor to play different types of music and sounds, experimenting with what attracted fish—Viennese waltzes, it turns out.[33]

As Hass's broadcasts illustrate, film soundtracks predominantly relied on sound created in air—dry made wet—rather than using underwater sound, and I mention soundtracks peripherally.[34] There are important exceptions, of course, such as the sound of breathing through the regulator, or dolphin vocalizations in *The Silent World* (ca. 14:04). Subsequently, dolphin vocalization, brought to public attention by John Lilly in the later 1950s, would play a role in enhancing the charisma of these mammals who were increasingly perceived as humans' sea counterparts; whale vocalizations would enter popular culture in the later 1960s with the humpback whale recordings made by biologists Roger and Katy Payne, and engineer Frank Watlington.[35] Sonar was a particularly important development in underwater sound, whose invention dated to World War II. Sonar figures in *The Silent World*'s search for a shipwreck, and it recurs throughout *Sea Hunt*. Sonar also plays an important role in films featuring submarines, which, however, are not studied here (with the exception of *20,000 Leagues*) because they contain few sequences in the underwater atmosphere—aside from an occasional image of the vessel's hull underwater, murky depths, or a looming enemy. Indeed, as I have mentioned, the obscurity of the depths is essential to submarine films' narrative and suspense, as sonar operators and crew, denied visual information, listen for the sounds signaling an approaching enemy.

2.2. *20,000 Leagues under the Sea*: Theatrical Realism

The "monster" disrupting nineteenth-century global shipping is not a creature but rather a futuristic submarine. Verne's Professor Aronnax, played by Paul Lukas, and his assistant, Conseil, given an anxious cast by Peter Lorre, make this discovery when they survive an attack on the US warship on which they are passengers. Stranded at sea, they clamber onto a platform that leads into the navigation room of a Victorian-era steampunk craft, filled with ostentatious instruments of measurement. Its atmosphere of nautical precision contrasts with the luxurious parlor Aronnax next steps into, which is bathed in a flickering blue-violet, almost otherworldly light. The light turns out to emanate from a punched-out picture window, an enormous porthole into the depths of the sea. A countershot from the outside shows a few green fish gliding by, but when we take the viewpoint of Aronnax and Conseil, we see an unexpected scene. Aronnax and Conseil gaze down at a procession of divers on the seafloor, clad in what at first glance appears to be helmet diving suits— yet their breathing apparatus is self-contained.

In the following entrancing sequence, less than two minutes long, we witness the divers moving in double file across the screen, carrying on a bier a prone humanlike form draped in white (fig. 2.2). Their movements are grave, suited to the mood of the scene and in keeping with the difficulty of moving through water, coordinated to mournful music. At the head of the procession is a figure bearing a cross encrusted with baroque undersea creatures. A brief ceremony follows; however, since the science fiction fantasy has not given

FIG. 2.2. Richard Fleischer, *20,000 Leagues under the Sea*, ca. 25:48. Funeral procession: box set conventions help define underwater space and create a sense of recession in the murky depths.

microphones to the divers, the leader conducts the ceremony with exaggerated gestures. The leader raises his hand in benediction, and divers kneel before a bier before they slowly troop back toward the submarine. Then the leader stops, and with another melodramatic gesture points toward Aronnax and Conseil, watching from the *Nautilus*. With a sweeping wave of his arm, the leader urges his divers on, and the scene shifts to the surface, where Aronnax, Conseil, and harpooner Ned Land (played by Kirk Douglas) will subsequently be taken prisoner.

Aronnax's exclamation explains the action: "a burial ceremony, under the sea" (ca. 25:40). The burial procession is the first view into the depths offered by Fleischer in *20,000 Leagues under the Sea* (1954), in a sequence that lasts a minute and a half. Magazines may have featured articles on the headaches of making the film, but there is no trace of difficulty in this masterful illusion. Using Technicolor and pioneering the wide-screen format of Cinemascope, *20,000 Leagues* won two Academy Awards in 1955 for best art direction and best special effects. The battle with the giant squid was singled out, but the film's artistry also included its organization of diffuse underwater space into tightly choreographed scenes.[36]

The previously mentioned strategy of familiarization is essential to this achievement. There are several techniques important to this strategy. As mentioned in part 1, the choice to take as frame Verne's novel is the first, for the novel *Twenty Thousand Leagues under the Seas* had endured as the most important script for imagining the undersea as a diverse planetary environment. The voiceover of Professor Aronnax, the narrator in the novel, further helps spectators match passages taken from the novel with the physical environment that Verne could only imagine but that the film is able to show.

Yet the most persuasive technique of familiarization is visual: to organize the underwater film set to resemble space as perceived on land. Part 1 detailed the ways in which underwater optics not only violate terrestrial optics with their drastically diminished visibility but also possess features inimical to Western techniques going back to the Renaissance for the realistic depiction of three-dimensional scenes on a two-dimensional plane. In the tradition of Pritchard, and then DeMille's wreck set, *20,000 Leagues* helped compensate for the low visibility and shallow depth of field by borrowing from proscenium theater the ability to create within compressed space settings capable of holding complicated actions among multiple characters.

The theatrical quality of the burial scene is evident from the moment when Aronnax and Conseil first discover the marvelous new world under the sea, which they watch from an elevated viewpoint. They watch through a giant window, a plausible terrestrial perspective on underwater life reaching back to the aquarium era of the later nineteenth century. When Paton invited viewers to look through Nemo's "magic window" in the 1916 *Twenty Thousand Leagues*, the viewer was on the same horizontal plane as the scene, or sometimes saw it tipped slightly downward. When Fleischer took up aquarium perspective in the Disney *20,000 Leagues*, he accentuated the spectator's view downward to about 30–45 degrees, as is visible in figure 2.1. This angle of viewing takes from proscenium theater the perspective of the choice box seats. It enhances a sense of depth by giving the eye a greater distance to traverse. It also offers the viewer the feeling of scopic mastery, giving the eye the ability to place elements in relation to one another in the field of action. Even as this angle is an apt aesthetic choice, it gave the camera operator a fixed position, overcoming the continuous movement of water.[37]

The divers' movement from left to right in their procession helps give the audience a sense of ease in the environment, as this direction feels natural to viewers from reading cultures with the Roman alphabet. The sense of ease is reinforced by the apparently spacious setting. The divers file by solemnly and naturally, as if they were walking on land, but the comments by professional diver Bill Strophal, one of the underwater stuntmen for the 1954 *20,000 Leagues*, expose the illusion. In fact, the divers were so tightly organized between the "two rows of coral" that they had difficulty executing the funeral rite in the screenplay. "[W]e were supposed to turn around to the side and pick up the coral and gently place it on the body but the only problem was there wasn't

FIG. 2.3. Mark Young, *The Making of 20,000 Leagues under the Sea*, ca. 41:46. Laying hemp carpets to tamp down the swirling sand stirred up by actors walking, which thwarted image capture.

enough room to turn around."[38] To enhance depth of field in such a confined space, the filmmakers borrowed from proscenium theater the use of flats on the set to create overlapping planes. These overlapping planes can give a sense of greater recession than in fact obtains, since it "is difficult for an audience to gauge the actual depth of the stage."[39] The flats in the burial scene are the environment's coral reefs, a natural feature that will be used for such effects going forward. With the planes of coral reefs suggesting recession, the film could harness the underwater haze, confusing the eye about the positioning of figures in relation to the viewer. In theater, comparable techniques include the use of smoke, scrims, and lighting. However, as Fleischer observed, "the sea is there to defeat you."[40] The turbid undersea had to be managed so it could throw a cloud over the action but not obscure it altogether. Among the difficulties in filming mentioned by the article in the 1955 *Picturegoer*, people on the set kicked up silt from the seabed, which became "a fog of sand."[41] Dive supervisor Fred Zendar found a solution: laying huge hemp carpets to keep the sand in place—not for long, with all the people walking, but as Fleischer observed, enough to get several takes (fig. 2.3).

The burial sequence further drew on the conventions of proscenium theater with its arrangement of bodies. The film integrates fanciful self-contained breathing apparatus into the helmet diving suits imagined by Verne, with steampunk diving gear now hiding scuba. Verne imagined divers weighted like the helmet divers of his time, using the familiar motions of walking, albeit on the seafloor rather than land—not with the fishlike ability to swim using all three dimensions of the underwater atmosphere enabled by scuba. With the floor of the stage serving as a ground plane for creating recession back to a vanishing point, divers can be placed on a slight diagonal to amplify the impression of receding space. The arrangement along diagonal lines to expand the viewer's sense of deep space is yet more pronounced than in the underwater burial scene when the *Nautilus*'s crew returns to the submarine in two diagonally arranged columns (fig. 2.4). The haze shrouds the vanishing point, masking its distance from us as well as the number of divers, whom we can imagine extending in great numbers to the horizon (in contrast to the view in the production still) (fig. 2.5). Another contrast that aids in grasping the film's artful production of terrestrial optics by importing stage conventions is the manta ray sequence that ends Hass's *Under the Red Sea* (1951), in which bodies loom and fade like the Cheshire cat in Lewis Carroll's *Alice in Wonderland* (fig. 2.6).

Fleischer's *20,000 Leagues under the Sea* returns to these techniques taken from theater in the next, more extended, underwater sequence, launched by the captives' visit to Nemo's extraordinary undersea cultivation (ca. 40:40). In rewriting Verne's hunt in the forest of Crespo to instead be a display of submarine cultivations, the film chooses a pretext for spectators to enjoy views of sea life, such as the majestic sea turtles, which are easier to film when they are constrained, than if the camera had to chase after creatures swimming freely (fig. 2.7). As this image illustrates, the divers holding the turtle's flippers come toward the viewers walking on a ground plane at a diagonal, while the coral reefs shrouded in haze confuse the eye about distance. The sequence uses similar techniques throughout the exhibition of underwater cultivation and harvesting necessary to sustain Nemo's renegade collective, which has vowed to abstain from products derived from land.

20,000 Leagues's episode of undersea cultivation ends with Land and Conseil sneaking away and discovering a sunken ship. In Verne's novel, the submarine comes across the wrecks of ships ranging from fragments of the eighteenth-century navigator La Pérouse's ships to a British vessel sunk by the *Nautilus*, which affords the gruesome vision of the newly drowned.[42] However, the characters in the novel do not exit the submarine to explore these

FIG. 2.4. Richard Fleischer, *20,000 Leagues under the Sea*, ca. 26:54. Divers returning to the *Nautilus* in diagonal columns to enhance viewers' perception of their numbers as well as a sense of space in the depths.

FIG. 2.5. Peter Stackpole, "*20,000 Leagues under the Sea*," photograph. The LIFE Picture Collection via Getty Images. View of the column from the perspective of divers involved in production.

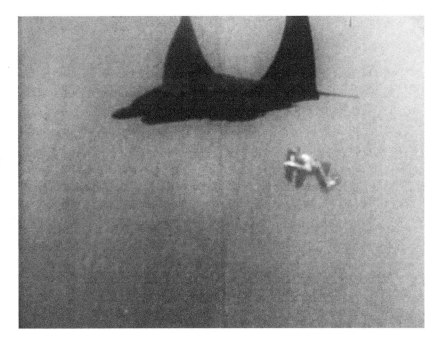

FIG. 2.6. Hans Hass, *Under the Red Sea*, ca. 59:46. A diver filming a ray: the undefined sense of distance true to underwater perception, in contrast to the box set conventions of *20,000 Leagues*.[43]

FIG. 2.7. Richard Fleischer, *20,000 Leagues under the Sea*, ca. 45:16. Nemo's underwater plantations. Exhibiting animals as captives keeps them still.

FIG. 2.8. Richard Fleischer, *20,000 Leagues under the Sea*, ca. 46:22. A spacious wreck set.

wrecks. The addition of this episode serves both a narrative and an aesthetic function. The narrative function is to launch the plot of a "jailbreak" led by Ned Land, played by Kirk Douglas, that was one of Fleischer's transformations of the novel when he adapted it to cinema. The aesthetic function is again to create space underwater scaled to the human body, enhancing the actors' ability to interact.

As we will see in Cousteau's films, a ship's wreckage is in fact a dangerous, crumbling site that the scuba diver swims through with caution. In the Disney *20,000 Leagues under the Sea*, in contrast, the wreckage is a roomy interior, which the set designers constructed according to the conventions of proscenium theater's box set. The backdrop is a ship's hull, whose parallelogram shape creates a sense that it curves back into the sea (fig. 2.8). Openings in the hull, whether ruins of windows or simply worn away by erosion, reinforce the connection to unshown action outside the wreck, integrating the set and spectator with the depths. The irregular, broken-down décor exudes an appropriate mood of mystery and contributes to the optical confusion, while the haze increases the sense of space in the wreck interior.

Land and Conseil have not been investigating for many seconds before the viewer glimpses through a dilapidated window a shark gliding by in the murky blue, unseen by the characters (fig. 2.9). If a pistol has been hung on the wall in the first act, it should be fired in the next one, runs the famous axiom by Chekhov. The same is true of sharks in the theater of the sea. The wreck scene will end with Nemo's discovery of Land and Conseil's escape, the shark attacking the renegades, and their rescue by Nemo. The shark attack

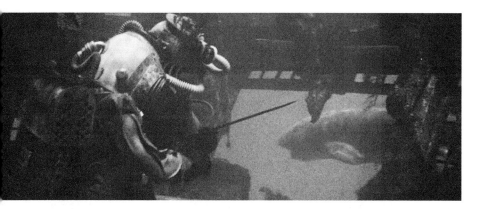

FIG. 2.9. Richard Fleischer, *20,000 Leagues under the Sea*, ca. 46:28. Glimpses of a shark through the wreck set window create a sense of deep space and danger.

is an exception among the adventure episodes of *20,000 Leagues*, which primarily show drama in the atmosphere of air and use the underwater film set chiefly for its novelty and mystery. Even so, the film was made at the beginning of an era when Verne's fantasy of divers untethered from the surface was fact. The possibilities of scuba could now be the subject of film, as well as instrumental to its production.

2.3. *The Silent World*: Creating a New Shot Vocabulary for the Aquatic Atmosphere

The Liquid Sky

The Silent World celebrates the vertical freedom of the human adventure beneath the sea with an exhilarating fifty-second cold opening, plunging us into the activity of diving.[44] In this sequence, divers glide smoothly down into the depths, as if they were flying in a liquid sky, breathing with scuba and propelled by swim fins. Almost every scuba pioneer, including Cousteau and Diolé in *The Undersea Adventure*, likened underwater movement to flying.[45] The streamlined, seemingly effortless trajectory of the scuba diver contrasts with the helmet diver's battle with the weight of the atmosphere, as the film will emphasize in a subsequent sequence contrasting the modern scuba diver as an icon of freedom with the old-fashioned helmet diver, shown at once as a brother and as a forebear.

FIG. 2.10. Jacques-Yves Cousteau and Louis Malle, *The Silent World*, ca. 0:19. Scuba divers as the modern Prometheus.

From the start, the film imbues the descent of the divers with a triumphal mood. Even before the credits, a white-orange flare is lit in deep blue water, in a harmonious contrast of color and form. The flare is shown to be one of a set carried by the team of divers, who descend with grace and focus, bearing the flares like torches with their muscular arms (fig. 2.10). Cousteau and Malle intercut the descent into the depths with bubbles rising up from the flares and the divers' regulators. The bubbles rise in clouds as the medium of water makes visible the invisible element that sustains life on land. The flare, the bubbles, and the divers' glide are as rich in provocative contrasts as they are beautiful. What kinds of torches burn in water? How does the atmosphere of water enable descent in a fashion other than falling, as an exhibition of power and control? Why do air bubbles rise with the spheres "peculiarly flattened like mushroom caps," to cite Cousteau's observation in the book *The Silent World*?[46]

In encouraging such questions, the opening scene of *The Silent World* coordinates the two strategies I have suggested as important for exhibiting unprecedented environments: at once to make audiences feel at home amidst its conditions and also to expand their imaginative horizons with its novelty. While *20,000 Leagues* turns to the long-standing tradition of the stage, *The Silent World* adapts the shot vocabulary of cinema, in keeping with Cousteau's commitment to celebrating the modernity of the undersea experience. With the initial plunge, the film, for example, submerges the convention of the establishing shot that reveals to spectators the setting where the action will unfold. One form of establishing shot on land is the crane shot, "commonly used to gradually reveal the grand scale of a location or environment as the camera is moved upward, including more details as the camera's vantage point gets higher."[47] To this point, I have emphasized the difficulty of working on the underwater film set, but Malle was also fascinated by its aesthetic opportunities: "The camera by definition—because we were underwater—had a mobility and fluidity; we could do incredibly complicated equivalents of what, on land, would be a combination of crane movements plus enormous tracking shots—and we could do it just like breathing because it was part of the movement of the diver."[48]

Even as Cousteau and Malle draw on the crane shot in the opening to *The Silent World*, they adapt it to the specific perspectives offered by the depths. Rather than expanding the field of view by gradually rising up, the camera remains in one zone at the surface and expands its field of view as it follows the divers down. When they reach their destination, the camera encompasses a setting that we will learn after the credits is a movie studio fifty meters beneath the sea. Fifty meters on land is not much distance; however, fifty meters beneath the sea is beyond the limits of the leisure diver, even today. The film conveys a sense of the significance of that depth with the length of screen time it accords to the divers descending (forty-five seconds, which, according to photographer Luis Marden, who dove with Cousteau during the making of the film, is precisely the time that flares last).[49] Nonetheless, despite the novelty of such a view, the conventions of the establishing shot familiar from land are well served by its conclusion. In having the divers reach the seafloor, the directors give the gaze a ground where it can come to rest, rather than letting our vision dissipate into the undefined haze.

In Fleischer's *20,000 Leagues*, the first view into the depths is of a solemn ceremony familiar on land: "a burial ceremony, under the sea." In a related fashion, the plunge of the divers opening *The Silent World* cites an iconic cultural pose. In showing divers descending with flares outstretched, Cousteau and Malle evoke a neoclassical icon for the advance of civilization, exemplified by the Statue of Liberty or the torch bearers of the modern Olympic games. Light illuminating the darkness is an emblem of knowledge—and it also is necessary for shooting underwater film. A classical figure igniting human innovation using fire is Prometheus, the Titan who stole it from the gods as a gift for humans. In Cousteau's subsequent *Window in the Sea*, almost twenty years later, he identifies French marine biologist Louis Boutan, the inventor of underwater photography, with Prometheus, who, in Cousteau's words, "had given mankind fire and thus artificial light."[50] In some well-known images, Prometheus takes fire down from Mount Olympus on foot, but in others, as in an image by Peter Paul Rubens (1636–37), he flies down through the air (fig. 2.11). Rubens's Prometheus looks over his shoulder furtively, expressing his sense of overstepping the boundary between humans and gods. No sense of transgression accompanies the divers' empowered descent. Although the dangers of the undersea will emerge repeatedly in the film, they are never suggested as divine retribution for human curiosity but rather are the reality of an inaccessible area of our planet. Even as it connects with tradition, *The Silent World*'s cold opening also intimates Cousteau's fantasy that humans would progress to aquatic existence. Crylen connects *The Silent World*'s sequences celebrating the freedom of scuba with a human metamorphosis into a creature Cousteau termed *homo aquaticus* in the early 1960s.[51]

While *The Silent World* is a documentary, note how the film diverges from documentary details in its opening sequence in service of creating such a myth. First of all, if the filmmakers had only lit the images with flares, the figures in the opening sequence would have "shown up only as . . . dark silhouette[s] had the photographer not used other lighting on it."[52] As Cousteau comments in *Window in the Sea*, "The flare with its thick stream of bubbles is beautiful in itself, but its use as a light source is somewhat limited."[53] Instead, the rich colors of *The Silent World* depended on the lighting designed for Cousteau by Harold Edgerton.[54] However, "cables streaming down from surface vessels or shore" would appear "bothersome" and encumber the scene.[55] A two-page spread from *The Ocean World of Jacques Cousteau* shows the type of cables used to bring down the immense amount of power needed for underwater lighting, left off screen in the opening sequence, although visible at other points in the film (fig. 2.12). Rather than documentary, the triumphal descent of the divers exemplifies Cousteau's "docudrama" criticized by Malle.

Another docudrama element in the opening sequence is the divers' nude torsos. The film dispenses with the wet suits used elsewhere, needed for insulation even in the comparatively warm seas where the film was shot. Cousteau notes that "[b]odily heat lost in sea bathing is enormous, placing a grave strain on the central heating plant of the body."[56] However, a wet suit would have

FIG. 2.11. Peter Paul Rubens, *Prometheus* (1636–37), oil on panel.
Museo del Prado.

attenuated the analogy between classical statue and diver—although the sequence instead has a strange mixture of the classical and the cyborg, given the
divers' masks and regulators that hide their faces. With their faces hidden and
unable to speak, they communicate with their bodies. The opening dive is a
water ballet, with duets and trios of men aligned to echo the parallel lines of
their air cylinders and of schools of fish. The medium shot that is an easy focal
point in the underwater haze exhibits their strength, and indeed the haze itself
adds glamour, as throughout underwater exhibitions of dance.

Problems of Filming

The aqualung freed us from the surface. But we still find ourselves tied to the surface for electrical power. Batteries work well for low power needs, but in filming with many powerful flood lights we require vast amounts of electricity. It was so in the past and it is

still true. In these photographs it is plain what the problem is: bothersome cables streaming down from surface vessels or shore. In one effort to keep cables under control we tried attaching buoys to keep them floating off to the side of the work area. Eventually all facilities underwater must be free of surface encumbrances.

60

FIG. 2.12. Jacques-Yves Cousteau, *The Ocean World of Jacques Cousteau: Window in the Sea*, 60–61. Cables needed for adequate underwater lighting that were hidden in the opening of *The Silent World*.

Malle does not comment on the choreographed quality of the opening scene in his recollections. However, such coordination must have taken work, if we accept comments about underwater choreography made by George Sidney, the director of aquamusical star Esther Williams's first film exhibiting underwater dance filmed in the depths, *Jupiter's Darling* (1955). Williams had already performed ballet steps underwater, with no breathing aid, in scenes such as the vaudeville performance at the Hippodrome, in *Million Dollar Mermaid* (1952) (ca. 1:13:12). However, this scene had been shot with a fixed camera in a tank. In *Jupiter's Darling*, in contrast, the production descended into the pool with the actress. With a new ability to capture the amplitude of dance in the depths, the film, as Sidney explained, "had to invent and develop steps and strokes. . . . Some of Esther's movements had to be tailored to the camera and vice versa. I wasn't dry for three months."[57] The photographs taken in the pool while the film was being made show the film crew working using scuba gear, while the extraordinary waterwoman-actress makes breath-holding look easy, maintaining her poised countenance, smiling, and keeping her eyes wide open throughout the shots (fig. 2.13).

FIG. 2.13. Detail from *Jupiter's Darling* contact sheet. LOOK Magazine Photograph Collection, Library of Congress, Prints & Photographs Division. Preparing to film in the pool set with Esther Williams, who makes underwater ballet breath-holding look easy.

In *The Silent World*, the triumphal mood of the opening scene recurs in a subsequent sequence showing scores of dolphins porpoising. In this sequence, Cousteau and Malle create that mood in their stitching together of topside and underwater shots, creating continuity across the waterline, as the dolphins plunge beneath the surface and then launch out of it again. Underwater as topside, moreover, we see a similar dense arrangement and coordination of dolphins together, almost like a peloton of cyclists.

Interspersed with such shots are views of the diver Laban, who descends into the submarine observation station, observing through glass in the comfort of air. There, he embodies the figure of the spectator in the movie theater watching dolphins swimming. In fact, action imagery of the dolphins beneath the surface was captured with scuba and underwater filming. Both Torma and Crylen note in this sequence a type of submarine traveling shot, which propels the camera through the aquatic atmosphere with gliding speed and mobility, achieving the sequence's expression of the dolphins' aquatic viewpoint.[58] The film, however, hides the scooters needed to capture such shots in the parade of the dolphins, keeping off-screen new technologies that could disrupt the viewers' pleasure in the sense of direct contact with wildlife.

Landscapes Sprung from a Nightmare

Along with celebrating the potential of *homo aquaticus, The Silent World* exhibits the undersea as a frontier of discovery and danger, continuing a tradition of maritime writing about surface exploration that was familiar to Cousteau, who came to diving from training in the French navy. Indeed, the prolix subtitle of the English version of the book *The Silent World* echoes titles from this literary lineage: *A Story of Undersea Discovery and Adventure, by the First Men to Swim at Record Depths with the Freedom of Fish*. Like the mariners in this literature of "discovery," "adventure," and exploration, the divers in both the book and film of *The Silent World* venture into uncharted spaces whose conditions take them and their technologies to their limits. The film first reveals the danger of such exploration when Laban, in pursuit of spiny lobsters, suffers from rapture of the deep, rips out his mouthpiece and nearly drowns. His buddy, Falco, sends him to the surface but neglects to accompany him, and Laban forgets to make a decompression stop.[59] According to Malle, this episode was shaped from a real incident in filming when a "diver suffers from *ivresse des profondeurs* [and] passes out under water."[60] While Malle wanted to include the blackout sequence, Cousteau suppressed it. Rather, Malle detailed, "When we got to the cutting room, Cousteau needed to beef it up, and, also, he wanted to explain." Laban's dive teaches audiences about nitrogen narcosis, decompression sickness, and the use of the decompression chamber, along with the importance of the buddy system.

The episode of Laban's near death is narrated retrospectively by Falco, from the deck of the *Calypso*, a setting infused throughout the film with an Enlightenment mood of optimism and control.[61] However, when Cousteau and Malle descend to the deep undersea rockface where Laban nearly dies, the film shifts to a darker tone. Throughout the episode about Laban and a subsequent sequence placed at the limits of unprotected diving, Cousteau and Malle submerge conventions from film noir to create a sense of the menace stalking a diver at the limits of compressed air. As they brilliantly grasp, film noir meshes well with exhibiting the zone where rapture of the depths occurs because it uses lighting, close-up, and framing suitable to underwater photography in these extremely low-light conditions, and film noir too portrays disturbed minds and fatal events.

Film noir is characterized by "imbalanced lighting, with a marked lack of fill light" (in contrast to the more unobtrusive "balanced mix of three-point lighting").[62] This unflattering light fosters a sense of peoples' violent personalities and twisted motives. A similar harsh lighting creates a sinister cast both to the episodes where Laban experiences rapture of the deep and when the divers take us to the greatest depths hitherto shown in the history of underwater film. In the depths beneath sixty meters, "brains are stalked by rapture of the deep," the narrator recounts of a dive to the limits of compressed air (ca. 19:50).[63]

Film noir's disorienting framing is another convention suited at once to the mobility of the underwater camera, the distorted experience of perception in the zone where divers are threatened by rapture of the deep, and its low-light conditions. Film noir is famous for "extreme high and low angles" in climactic, violent sequences. The portable underwater camera easily accesses such extreme angles. The need of the underwater photographer to shoot close to the subject to reveal its contours in dim lighting fits as well with film noir's "'choker' (very tight) close-ups, and visual distortion." Such shots are terrifying in the scene where Laban almost drowns. The sequence shows us Laban's hand grotesquely magnified into a monstrous appendage as it plays with seaweed, becoming something alien to him and eventually, when the hand removes his regulator, threatening his survival (fig. 2.14). In the subsequent

FIG. 2.14. Jacques-Yves Cousteau and Louis Malle, *The Silent World*, ca. 11:07. The diver's hand becomes "abject," neither subject nor object, steeped in a horror palette, in this zone of "rapture of the deep."

episode taking us to the limits of scuba, Cousteau and Malle similarly use such close-ups to disorient the spectator, although the plot does not threaten a diver. Thus, this screen grab of seaweed removes contextual information that might give us a sense of scale or enable us to identify the type of organism being shown, confronting us with the bizarre quality of marine life, whether flora or fauna (fig. 2.15).

In order to undermine spectators' confidence that they grasp the whole picture intellectually as well as perceptually, one technique of film noir is to use "foreground obstructions or . . . shadows" that contrast with "an open, unobscured view of subject matter." Such a view meshes well with the shallow depth of field in underwater darkness. Thus, in the sequence where Laban suffers from nitrogen narcosis, shots of rocks block our access to the entire scene. Similarly, when Cousteau and Malle show the reef in the second scene at great depth, there is no overarching view to enable spectators to establish a relation among elements shown. Rather than offering an

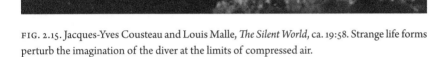

FIG. 2.15. Jacques-Yves Cousteau and Louis Malle, *The Silent World*, ca. 19:58. Strange life forms perturb the imagination of the diver at the limits of compressed air.

establishing shot, Cousteau and Malle montage disconnected views in jumpy fashion, echoing the disjunctive editing used in scenes of paroxysmal violence in film noir.

Film noir was typically shot in black and white, while one of the glories of *The Silent World* is its ability to exhibit the undersea in vibrant color. Yet the two episodes at great depth bleach out color, evoking film noir's lugubrious palette. As is visible in figure 2.14, the harsh lighting gives Laban's skin and wet suit a bleached-out, skeletal tone. To its blacks and pallid whites and off-whites the film adds reds, pinks, and browns. In showing red, the filmmakers are celebrating new technologies and their capacity to contribute new knowledge about the underwater environment. Thanks to the Edgerton strobe, film was able reveal "a vivid pigmentation never seen before even by the resident fish," in the words of the *National Geographic* caption accompanying one of Cousteau's first widely disseminated photos of red underwater, published in 1952.[64] At the same time, the voiceover tells us, "the more they descend, the more the browns become warm, the more the reds become violent," reinforcing the

sense of menace (ca. 19:36).[65] Such a sinister mood contrasts with the celebration of red in Hass's film *Under the Caribbean* (1954), which was finished
before *The Silent World*. *Under the Caribbean* builds to the exhibition of red
with lyrical language promising revelation: "Under the blue veils of the deep,
nature conceals a truth that we are anxious to discover. Colors never yet exposed to light will be revealed by our lens" (ca. 43:30). The sequence shows
divers taking down huge, strong lights that beam yellow amidst the ambient
blue. The revelation of red in the depths occurs when the camera approaches
a light illuminating a rockface, and its seeming blue turns red, in a sequence
exhibiting the power of technology rather than the disturbed psychology of
nitrogen narcosis.

Marine life at two hundred feet, the depth where Laban almost drowned,
is depicted in a similarly lyrical fashion in a photograph taken by *National
Geographic* staff photographer Luis Marden, shot when he followed Malle,
Cousteau, and the divers and seamen making *The Silent World*. In the photo
captioned "Through Caverns Measureless to Man, Down to a Sunless Sea,"
Marden alludes to Coleridge's opium dream, "Kubla Khan," where a drug trip
resonates with rapture of the deep (fig. 2.16). At the same time, Marden's portrayal glows in keeping with his Romantic citation. The luminous fish floating
calmly against the deep-sea twilight rhyme with the angle of the rock outcrop-
ping that anchors the shot in the right foreground. Marden organizes even the
white sea whips in a line along the slope of a rock, giving harmony to a "low-
order animal," to cite the photo's caption, that might otherwise seem creepy.[66]

In the Marden photograph, we can identify fish as well as the strange sea
whips, and Hass includes fish at these depths as well. Cousteau and Malle, in
contrast, fill their landscapes of nightmare with indeterminate forms. Our difficulty in identifying to what order these creatures belong recalls Bataille's disturbing *informe* used in surrealist fantasies. Cousteau had already suggested the
abject quality of deep-sea life in his book *The Silent World*, when he described
"vertical reefs two hundred feet deep," the same depth as the Marden photo.[67]
He described the growth on these reefs with comparisons to squishy forms,
whether sinister (tumors), life-giving (udders), or possibly both (mushrooms):
"on the rocks are living tumors and growths resembling udders, long fleshy
threads, chalice-shaped formations, and forms like mushrooms."[68] Marden's pale
sea whips similarly could have looked like a creepy, bizarre grove had he not
framed a view where they intermingle harmoniously with glowing fish.

Throughout much of the book *The Silent World*, as throughout the film,
Cousteau is eager to demystify pervasive fears of the deep. He takes aim, for

FIG. 2.16. Luis Marden, "Through Caverns Measureless to Man, Down to a Sunless Sea," in "Camera under the Sea," *National Geographic*, February 1956. Marden's photograph, accompanied by a caption from Coleridge's "Kubla Khan," brings glow and mystery to this same zone that Cousteau expresses in horror tones.

example, at the fearsome portrayal of the octopus in Victor Hugo's *The Toilers of the Sea*, where Hugo suggests that the worst death is to be "drunk alive" (Cousteau's wording) by octopus suction. In fact, the divers of the *Calypso* discover that the octopus is a shy, evasive creature. Cousteau remarked, "After meeting a few octopi, we concluded that it was more likely that to be 'drunk alive' referred to the condition of the novelist when he penned the passage, than to the situation of a human meeting an octopus."[69] Within such an overarching Enlightenment outlook, the deep rockface is an exception. The narrator of the film *The Silent World* ties its horror to the threat to divers at the limits of diving with compressed air. In concluding the film noir montage showing this environment, the narrator cautions, "One must not go deeper. One should not stay longer" (ca. 20:22).[70]

When *The Silent World* takes us to depths beyond direct human access, in contrast, it evokes wonder rather than horror. Photos achieved remotely by the *Calypso* at three hundred meters deep, using a camera let down over the ship's side, reveal "groups of organisms with strange forms," resembling "a starry sky," shown against the background of darkness (ca. 18:23; ca. 18:18).[71] The team of the *Calypso* contemplates this heavenly imagery found in the depths of submarine night, as the soundtrack's delicate music intimates the wonders of the physical cosmos made visible through remote photography.

The Gothic Wreck

Cousteau's first film shot with scuba was *Sunken Ships* (*Epaves*), and his attraction to wreck sites continued in the 1950s. He wrote lyrically about his experiences wreck diving in *The Silent World*, calling the wreckage of the four thousand–ton freighter *Le Tozeur* "a fine movie studio." According to Cousteau's description, it resembled the setting of a gothic novel, favoring ruined abbeys and castles. Cousteau characterized the ship as "an iron abbey," with "tunnels" and an "intricate stairwell." *Le Tozeur* had "bulkhead openings . . . arched in the fashion of cloisters" and "light filtering down as though from clerestory windows." This hazy atmosphere of mystery was ornamented by "[w]eeds that grew like lichens in a damp chapel," as marine life had taken over once useful technologies. As is common in gothic settings, Cousteau noted the chiaroscuro creating an atmosphere of mystery and menace. Divers could look from "a shadowy gangway at remote blue lights opening on the sea," whose danger Cousteau indicates with the comment, "we did not feel ready to swim through the tunnels to these lights."[72]

Cousteau's citation of gothic conventions in both *Sunken Ships* and *The Silent World* extends from the atmosphere of the wreck site to the narrative framing its exploration. The gothic novel, invented in the eighteenth century, takes readers into a haunted world after assuring them that, in their own Enlightenment era, the barbarism imagined, often of the medieval past, no longer obtains. *Sunken Ships* uses a topside prelude to explain dive and film technology; in *The Silent World*, the topside prelude shows the systematic processes used by the *Calypso* to locate the wreckage, scouring the area where the ship is known to have sunk and marking depths with the aid of sonar. The crisp, methodical search unfolds under the bright sun shining on the deck of the marine research vessel. A reminder of modern technology pervades even the soundtrack of the film in the charting sequence, accompanied by the hum of the ship's motor and punctuated by a ticking clock.

This mood of optimism could conceivably have extended to the portrayal of the wreck, as it was an immense achievement to locate a shipwreck and bring it to light; further, the location of the wreck awakens it to the present. In *Sunken Ships*, the wreck is lifted from the depths and melted down. In *The Silent World*, Cousteau and his team show how scuba and film propel the development of nautical archaeology. Chiseling off the overgrown bell of the wreck, Cousteau and his divers identify the ship as the SS *Thistlegorm*, a merchant ship built in Britain in 1940. The ship transported refurbished wartime materials to British troops in Egypt and Libya during World War II, until it was sunk in 1941 by a German airstrike in the Red Sea off the coast of Egypt. Since the time when Cousteau brought some of the *Thistlegorm*'s artifacts to the surface, the ship has become one of the most dived wrecks in the world. In In the words of a dive website, it is "a veritable underwater 'World War II Museum.'"[73] The wreck's accessibility would further expand in 2017 with the creation by its archaeologists of a virtual website allowing a worldwide public to tour it remotely.[74]

However, an ominous mood displaces such optimism when Cousteau's divers descend to explore the *Thistlegorm*'s wreckage, provoking spectator curiosity about what it will contain, along with apprehension about its possible perils. Such an atmosphere recalls the atmosphere created when gothic literature takes viewers from a present ruled by reason to a tortured, haunted past. Comparable to the gothic details Cousteau gives about *Le Tozeur* in his book *The Silent World* is, for example, Ann Radcliffe's description in *The Mysteries of Udolpho* (1794) of her heroine's nighttime exploration of a medieval fortress with "almost roofless walls." In this setting, the heroine, Emily, notes "the gothic points of the windows, where the ivy and the briony had long supplied the place of glass." Wind whistles through "narrow cavities," including winding staircases, vaults, and narrow passages "dropping with unwholesome dews." Such architecture confuses Emily and adds to her apprehension about the violence and ghosts lurking in this "obscure and terrible place."[75]

In sequences showing wreck sites in both *Sunken Ships* and *The Silent World*, Cousteau and Malle find a cinematic equivalent to such gothic language, making use of the new shot possibilities offered by the physical qualities of the

FIG. 2.17. Jacques-Yves Cousteau and Louis Malle, *The Silent World*, ca. 40:16. The wreck has settled on its side on the seafloor, tilting as if shown with a canted shot.

aquatic environment. The wrecks that are the focus of both films lie off-kilter on the ocean floor, and the camera uses slanted camera angles throughout their exhibition. The portrayal is particularly dramatic in the lush color of *The Silent World* (fig. 2.17). In cinema shot in air, a markedly off-kilter shot is called a "dutch angle" or canted shot, "composed with a camera tilted laterally, so that the horizon is not level and vertical lines run diagonally across the frame."[76] Gustavo Mercado credits the invention of this shot to German Expressionism, which destabilized rectilinear orientation to suggest "dramatic tension, psychological instability, confusion, madness, or drug-induced psychosis." As developed in subsequent terrestrial cinema, the shot is used to portray disturbance of both characters and settings. In the sequence of diving near the *Thistlegorm*, Cousteau and Malle mix off-kilter views that are an accurate representation of how the wreck settled on the ocean floor with tilted shots that capture the multiple orientations of the diver's perspective as he explores the wreck in a three-dimensional atmosphere. These shots are further equated with a human perspective because they look handheld, shaking subtly. Even if the diver is in control, such views convey apprehension and disorientation, in

FIG. 2.18. Jacques-Yves Cousteau and Louis Malle, *The Silent World*, ca. 43:21. A fish at an absurd window, now separating sea from more sea.

keeping with the mood of visiting a long-veiled grave site overgrown with algae, which has become the dwelling place of fish.

Images of terrestrial elements stripped of their use below add to the eerie atmosphere of the wreck sequence in *The Silent World*. At one moment, we see fish looking through what was once a window onto other fish beyond (fig. 2.18). However, windows and portholes are absurd in an environment where the distinction between inside and outside has been erased. Another example of such an image is the glimpses of a coil of rope dangling from the edge of the ship, as shown in figure 2.17. A basic, multipurpose nautical technology, the rope here has descended to Davy Jones's locker in a shape that resembles a hangman's noose.

The Silent World's undersea gothic also transforms the traditional portrayal of the gothic spectator who traverses landscapes haunted by the past. In gothic novels like *The Mysteries of Udolpho*, readers identify with Emily's lively curiosity as she explores ruins, sharing her fear and determination to comprehend and demystify the past. While viewers identify with the crew of the *Calypso* as they search for the wreck and share their excitement in its discovery, the figure of the diver becomes more enigmatic beneath the surface, notably in *The Silent World*.

Sunken Ships uses a voiceover to describe the actions of the diver who "glides without effort" (ca. 7:49), and the areas of the wreck that he visits. The diver's smooth gliding is, of course, practical: a hallmark of efficient swimming that further helps the diver avoid contact with the dangerous environment. In the book *The Silent World*, Cousteau remarked that *Le Tozeur* "was as treacherous as she was receptive, and taught us much about the actual perils of wrecks," from the risk it would collapse to organisms living there that could injure the diver.[77] Adding to the visual impact of such imagery is the notion that ghosts are reputed for their gliding movement, and the voiceover in *Sunken Ships* emphasizes the diver's presence on the wreck—remarking of the air bubbles from the regulator that linger in its spaces, "we leave behind us phantom beings [un peuple de fantômes]" (ca. 10:46). In *The Silent World*, the voiceover disappears, and spectators are left to make their own connections between the gliding diver exploring the wreck and the ghosts of people who once used these spaces. The music introducing the sequence stops at its apex, when the diver explores the wreck's compartments now inhabited by fish (ca. 42:26). His presence is then rendered yet more uncanny, offering an image that joins the human and the haunter, intensified by the harsh sounds of his breathing through the regulator.[78]

While Verne imagines the faces of the drowned when he portrays shipwreck in *Twenty Thousand Leagues*, such horror is eschewed in Cousteau's exhibition of the wreck as gothic spectacle.[79] This stance may be in keeping with Cousteau's exaltation of the new human access to the undersea, which dedramatizes its dangers. Cousteau's stance as demystifier may also explain the appeal for him of gothic spectacle as a way to organize wreckage. The gothic imagines ghosts as aesthetic phenomena, using poetic devices to at once evoke them yet hold them at bay so that their stories can enrich but not overwhelm the present. The aesthetic approach to the dead is evident in the visit to the *Thistlegorm*, where we see only melancholy traces of humanity, like the useless window and rope, an empty boot, or toppled lanterns.

Gothic novels offer charged moments at the climax of the drama when the dead spring to life—whether in the imagination of the observer or, when these narratives take a supernatural turn, in the form of ghosts. Such an electrifying moment of contact between the present and the past is found in Cousteau's underwater gothic as well. In *Sunken Ships*, there are two moments of contact: the diver grasping the iconic eight-spoked wheel that, the voiceover tells us, "[a]n old helmsman took ahold of . . . for its first launch," as well as finding a tube that "was a cannon. The wheel used to aim it still turns" (ca. 8:04).[80]

Plate 1 (fig. 0.3) William Lionel Wyllie, *Davy Jones's Locker* (1890), oil on canvas. Photograph © National Maritime Museum, Greenwich, London. Wyllie has captured the submarine palette and haze, yet the clearly defined detail in the foreground of the scene, such as the octopus entwined with human remains, characterizes seeing through air rather than water.

Plate 2 (fig. 1.1) "The Aesop Prawn, &c." from Philip Henry Gosse, *The Aquarium: An Unveiling of the Wonders of the Deep Sea* (London: John Van Voorst, 1954), plate 6, p. 222. Photograph © The Royal Society. Gosse's installation frames this cold-water shrimp native to British waters, *Pandalus montagui*, as an exquisite specimen in a miniature baroque grotto.

Plate 3 (fig. 1.4) Eugen von Ransonnet-Villez, "Coral Group in the Harbor of Tor." Photograph © Natural History Museum, Vienna. The image is used in Ransonnet's *Reise von Kairo nach Tor zu den Korallenbänken des rothen Meeres* (Vienna: Carl Ueberreuter, 1863), fig. 1, p. 35, where it is described as follows: "An outstanding group of corals, like an oasis out of the muddy sand, in a place that is protected from the waves. During low tide, the seafloor is covered by about six feet of water." The forms and colors of the scene convey aquatic perspective, yet Ransonnet also frames and flattens the scene in keeping with the conventions of the aquarium.

Light penetration in open ocean **Light penetration in coastal waters**

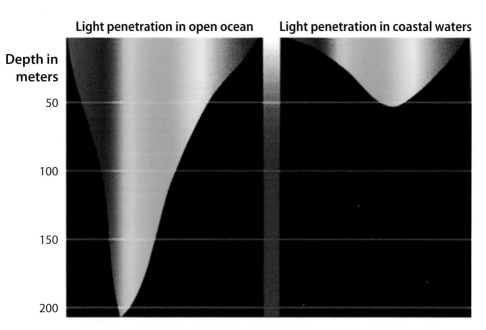

Depth in
meters

50

100

150

200

Plate 4 (fig. 1.3) "Light Penetration in Open Ocean / Light Penetration in Coastal Waters." Kyle Carothers, NOAA-OE. The effect of depth on human perception of color underwater, when unaided by artificial light.

Plate 5 (fig. 1.5) Eugen von Ransonnet-Villez, "Two Groups of Madrepores off the Coast of the Island of Ceylon [present-day Sri Lanka] near Point de Galle Drawn from a Diving Bell, February 3, 1865" (Deux groupes de Madrépores sur la côte de l'Ile de Ceylan près de Point de Galle dessinés dans une clôche à plongeur. 3 février, 1865), oil on paper. Photograph © Institut océanographique, Fondation Albert 1er, Prince de Monaco. Corals loom in three dimensions toward the viewers in this exquisite scene drawn from a diving bell, conveying the enveloping sense of the liquid atmosphere.

Plate 6 (fig. 1.6) Walter Howlison "Zarh" Pritchard, *Parrot Fish and Poisson d'Or amongst the Coral in the Lagoon of Papara Tahiti* (ca. 1910), oil on leather. Courtesy of the Scripps Institution of Oceanography, UC San Diego. According to Pritchard, at a depth of about thirty feet, where he painted this fish (a yellowtail coris), the unaided human eye perceived coloring in the "higher, delicate tints of pale greens, grays, and yellows."

Plate 7 (fig. 1.7) François Libert, "Yellowtail Coris, Terminal Phase—*Coris gaimard*," photograph. When viewed with the help of artificial lighting, the coris's bright colors dazzle undersea, as in this contemporary photo of a coris at the same stage of development as the coris painted by Pritchard.

Plate 8 (fig. 1.26) Cecil B. DeMille, *Reap the Wild Wind*, ca. 1:53:04. Artifice delights viewers in this scene of combat with an obviously fabricated giant squid.

Plate 9 (fig. 2.2) Richard Fleischer, *20,000 Leagues under the Sea*, ca. 25:48. Funeral procession: box set conventions help define underwater space and create a sense of recession in the murky depths.

Plate 10 (fig. 2.10) Jacques-Yves Cousteau and Louis Malle, *The Silent World*, ca. 0:19. Scuba divers as the modern Prometheus.

Problems of Filming

The aqualung freed us from the surface. But we still find ourselves tied to the surface for electrical power. Batteries work well for low power needs, but in filming with many powerful flood lights we require vast amounts of electricity. It was so in the past and it is

still true. In these photographs it is plain what the problem is: bothersome cables streaming down from surface vessels or shore. In one effort to keep cables under control we tried attaching buoys to keep them floating off to the side of the work area. Eventually all facilities underwater must be free of surface encumbrances.

60

Plate 11 (fig. 2.12) Jacques-Yves Cousteau, *The Ocean World of Jacques Cousteau: Window in the Sea*, 60–61. Cables needed for adequate underwater lighting that were hidden in the opening of *The Silent World*.

Plate 12 (fig. 2.14) Jacques-Yves Cousteau and Louis Malle, *The Silent World*, ca. 11:07. The diver's hand becomes "abject," neither subject nor object, steeped in a horror palette, in this zone of "rapture of the deep."

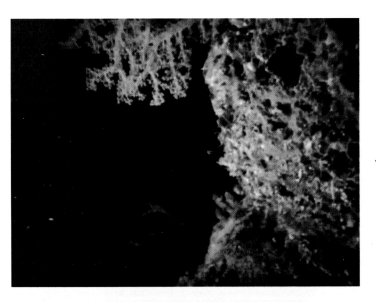

Plate 13 (fig. 2.15) Jacques-Yves Cousteau and Louis Malle, *The Silent World*, ca. 19:58. Strange life forms perturb the imagination of the diver at the limits of compressed air.

Plate 14 (fig. 2.16) Luis Marden, "Through Caverns Measureless to Man, Down to a Sunless Sea," in "Camera under the Sea," *National Geographic*, February 1956. Marden's photograph, accompanied by a caption from Coleridge's "Kubla Khan," brings glow and mystery to this same zone that Cousteau expresses in horror tones.

Plate 15 (fig. 2.19) Jacques-Yves Cousteau and Louis Malle, *The Silent World*, ca. 55:51. Underwater view of sharks feeding: predation.

Plate 16 (fig. 2.21) Jacques-Yves Cousteau and Louis Malle, *The Silent World*, ca. 56:01. Topside (surface) view of sharks feeding: white water chaos.

Plate 17 (fig. 3.3) James B. Clark, *Flipper*, ca. 56:32. The liquid perspective of Flipper.

Plate 18 (fig. 3.17) James Cameron, *The Abyss*, ca. 1:07:32. The liquid perspective of an NTI.

Plate 19 (fig. 3.22) Luc Besson, *The Big Blue*, ca. 59:09. Social pretensions displaced: Jacques and Enzo drinking *in* the pool.

Plate 20 (fig. 3.28) Michael Bay, *Pearl Harbor*, ca. 1:42:25. The viewpoint of the drowning sailor.

Plate 21 (fig. 3.31) Kevin Reynolds, *Waterworld*, ca. 1:24:38. The sunken Atlantis of modernity, aka "Dryland."

Plate 22 (fig. 3.36) Wes Anderson, *The Life Aquatic with Steve Zissou*, ca. 2:33. Steve Zissou with a school of fluorescent snappers at the premier of a film-within-a-film.

Plate 23 (fig. 3.40) "On the Underwater Filmset of *The Life Aquatic*." Photograph courtesy of Hydroflex, Inc. Underwater director of photography, Pete Romano, filming amidst the set's exuberantly fake seaweed in the episode "Investigating the Phantom Signal."

Plate 24 (fig. 3.37) Wes Anderson, *The Life Aquatic with Steve Zissou*, ca. 1:47:48. The jaguar shark's jaws from the spectators' point of view.

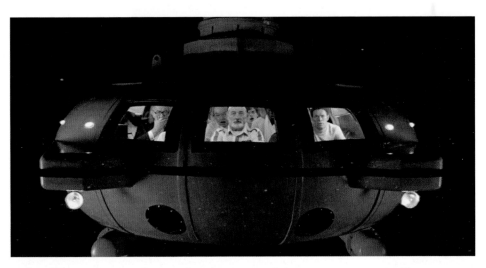

Plate 25 (fig. 3.38) Wes Anderson, *The Life Aquatic with Steve Zissou*, ca. 1:47:31. The spectators from the jaguar shark's point of view.

Plate 26 (fig. 4.2) James Honeyborne and Mike Brownlow, producers, *Blue Planet II*, episode 1, "One Ocean," ca. 22:17. The mobula ray's "extraordinary ballet of life and death" (ca. 21:46).

Plate 27 (fig. 4.3) Terence Young, *Thunderball*, ca. 4:53. Maurice Binder's credits mimic a bioluminescent dance of hunter and hunted.

Plate 28 (fig. 4.4) James Honeyborne and Mike Brownlow, producers, *Blue Planet II*, episode 6, "Coasts," ca. 9:03. Shot using the megadome, which can focus at once above and below water, in this view of a sea lion hunting tuna.

Without the voiceover, the contact between present and past is yet more enigmatic in *The Silent World*. When the diver finds the ship's bell, before reading
the inscription, he strikes this meaningful artifact twice, awakening it from its
long sleep. Only then does a female voice read the legend on the bell: "*Thistlegorm* Glasgow" (ca. 37:39). Ships' bells, which nautical archeologists use to
identify vessels, are "the 'symbolic embodiment of the ship itself,' . . . used to
mark the passage of time onboard the vessel, struck every half-hour, day or
night, as well as to signal the change of the crew's watches."[81]

In a recent article on gothic temporality, Jesse Molesworth asks why gothic
fiction, set in the religious past, is fascinated with modern clock time. He singles out notably the genre's fondness for the fatal hour when dire events occur.
In Molesworth's explanation, the gothic preoccupation with the hour offers a
kind of sacred time for a secular age. The homogeneous hours of the clock
become qualitative: "not mere witnesses" to plot but rather "each possess[ing]
an almost talismanic power."[82] Molesworth's point is relevant to the moment
when the diver strikes the bell of the *Thistlegorm*. The bell here has a talismanic
knell, recalling the dead lost beneath the sea, as well as signaling the arrival of
the diver summoning the wreck into his present. It reverberates through air,
at once recalling and contrasting with the sound of a ticking clock, which we
previously heard accompanying the topside work of locating the wreck. Similarly talismanic is the female voice that then identifies the ship, reading text on
the bell with a British accent, which is not connected to a character in this film
entirely peopled with men and narrated in French. Nor was a woman among
the *Thistlegorm*'s crew or the nine who died there. Her disembodied voice
speaks through air as well. Such sounds from the terrestrial atmosphere accentuate our ambivalent sense of human presence beneath the sea in the *Thistlegorm* sequence, in contrast to other episodes in the film.

In yet one more emotional nuance, Cousteau and Malle's sequences of
wreckage intimate new forms of life that differ from the historical contributions of nautical archaeology. Cousteau struggled to put into words this ambivalent sense of potential emerging from wreck diving in the book *The Silent
World*, echoing the narration of creative contact between present and past in
Sunken Ships: "To one who glided easily across the moss-covered deck, nothing was wood, bronze or iron. The ship's fittings lost their meaning. Here
was a strange tubular hedge, as though trimmed by a fancy gardener. Didi
[Frédéric Dumas] reached into the hedge and turned a wheel. The cylinder
rose smoothly. It was a gun barrel."[83] The energy of such a strange amalgam,
artistic hedge and gun barrel, recalls surrealism and its aesthetic of convulsive
beauty. In *The Silent World*, as well, Cousteau and Malle offer visual

expressions for the wreck's new potential more in keeping with surrealism, or perhaps new sea ontologies. In one more uncanny moment, we watch the diver's bubbles escaping from inside the wreck through openings on the deck, reminding us the diver is well, but also perhaps as if the sunken ship itself were breathing. The camera then follows one of these bubbles floating upward, expanding like a mushroom as it rises into lighter atmosphere, in an eerie echo of the bubbles floating upward in the film's opening triumphal dive.

Shark Carnage

Cousteau invented the shark cage, which enabled the divers of *The Silent World* to descend among sharks and capture their feeding frenzy. The film introduces this episode by drawing on age-old clichés. "Sailors the world over hate sharks," declares the narrator (ca. 56:14).[84] The sharks in the film are frightening as they lunge toward the camera and savagely tear at the carcass of a baby toothed whale (*cachalot*) that was accidentally killed by the propellers of the *Calypso* when the ship placed itself in the midst of a pod of whales. Cousteau and Malle may have been inspired by Hemingway's portrayal of sharks in *The Old Man and the Sea* (1952), as this novel was then at the forefront of public attention, cited in Hemingway's Nobel Prize award in 1954. In any case, the mood of the scene could come straight out of the shark carnage imagined in the novel, although the melodramatic narration violates Hemingway's understated style: "the hellish circling of sharks around their prey—if some men have seen it, they have not lived to tell the tale" (ca. 53:39).[85]

If the narration is clichéd, the cinematic language is novel. This episode adapts to the underwater environment the shot/countershot pattern found in horror sequences of violence, showing the perspective of the killer and of the prey. This pattern dates at least as far back as F. W. Murnau's *Nosferatu* (1922). In the shark episode, Cousteau and Malle meld the alternating points of view of killer and prey with the limits of topside visibility and the ability of the camera to exhibit violence beneath the surface. The film first shows us seamen peering into the water trying to make out the action of the sharks. Spectators then become privy to the underwater view that the cameramen are able to capture (with the protection of the shark cage): sharks tearing at the carcass in what the film frames as savage predation (fig. 2.19). With the view of the carcass from below, further, we are given the perspective of the shark, whose eye the film also displays, encouraging us to look with the cold gaze of the predator (fig. 2.20). We then are shown the topside action, including the shark's

FIG. 2.19. Jacques-Yves Cousteau and Louis Malle, *The Silent World*, ca. 55:51. Underwater view of sharks feeding: predation.

FIG. 2.20. Jacques-Yves Cousteau and Louis Malle, *The Silent World*, ca. 54:03. The predator's eye.

FIG. 2.21. Jacques-Yves Cousteau and Louis Malle, *The Silent World*, ca. 56:01. Topside (surface) view of sharks feeding: white water chaos.

distinctive dorsal fin, chunks taken out of the carcass, and white, churning water stained red (fig. 2.21).

Other conventions adapted from horror movies in the portrayal of shark carnage include jerky camera movements, out-of-focus or dramatically tilted shots and abrupt cuts. To these, *The Silent World* adds the underwater haze that creates a distinctive watery palate of violence. Gray and white indicate the sharks feeding; the whale carcass is black with white flesh; and the predation is a swirl of blue water, white froth, and clouds of red blood.

Cousteau and Malle's melodramatic shark carnage is as much an aesthetic construction as the Promethean descent of the divers. The most prominent contrast in this era was Hans Hass's representations of sharks in *Humans among Sharks* (*Menschen unter Haien*) (1947/48), a black-and-white film made using free diving and a helmet without a suit in tropical waters. A pioneer in shark filming, Hass represents sharks as apex predators occupying a natural place in the food chain of the ocean. The divers film sharks calmly, on the alert, to be sure, but without melodrama, remarking on their grace and majesty. Cousteau would go on to revise the savage mythmaking of *The Silent World*, subsequently filming divers working in the presence of sharks. But the film's poetics of shark carnage

FIG. 2.22. Peter Gimbel and James Lipscomb, *Blue Water White Death*, ca. 43:34. "Action . . . as poetic as anything I've seen on the screen in a long, long, time" (Canby, "Screen: Dramatic Pursuit," 15).

went on to an imaginative life in cinema far outpacing reality. Just two years after *The Silent World*, *The Old Man and the Sea* (1958), directed by John Sturges, drew on the topside/underwater shot contrast and the use of color to show sharks devouring the fisherman's majestic marlin. It is hard to imagine that *The Silent World* was not seen by Boren, who shot the underwater sequences in *The Old Man and the Sea*. As I discuss in part 3, Peter Gimbel and James Lipscomb's *Blue Water White Death* (1971) and then Steven Spielberg's *Jaws* (1975) amplified Cousteau and Malle's portrayal of shark violence to mythic proportion (fig. 2.22).

Fish-Eye Perspective

Both Crylen and Torma underscore *The Silent World*'s innovations in creating for us shots that convey the perspective of marine creatures. Crylen emphasizes the scooters that enable divers to move with the grace and speed evident in the parade of dolphins. Torma underscores the film's use of the suggestively named wide-angle "fish-eye" lens, which "allowed 180-degree shots of the underwater world."[86] Whether or not such a lens approximates the vision of fish, Cousteau and Malle certainly give us a shark's point of view on underwater action in the shark episode, when the camera inspects or lunges at the carcass. The most extensive sequence suggesting a fish's perspective is the lyrical tour of a tropical reef toward the end of the film. The focus of the camera ranges from close-ups of a moray eel in its lair, or a parrot fish nibbling on coral, to graceful, fluid traveling shots and panoramas of schools of colorful fish, in one of the most prolonged sequences of the film without humans in the frame.

It is intriguing to simulate a fish's point of view, but could a human hope to enter into a fish's mind? At moments, *The Silent World* anthropomorphizes marine life. The baby whale that runs into the *Calypso*'s propellers, is "careless like any kid," and, after it is mortally wounded, "struggles courageously to rejoin its family," and swims a few moments with them before the crew of the *Calypso* give it the coup de grâce (ca. 50:08; ca. 51:13).[87] When researchers use dynamite to catch fish in another episode about the superiority of dive filming to inspecting dead specimens for knowledge of fish, the camera lingers for almost a minute on a blowfish, in the atmosphere of air, that has puffed up, but deflates, expelling water, as it gasps and dies. "The trick is powerless against dynamite," declares the narrator, accompanied by wistful music (ca. 24:02) (fig. 2.23).[88]

The closing sequence on the reef, punctuated with shots suggesting fish points of view, concludes with a question about the limits of such anthropomorphism. The human interest in the episode is "the friendship between a man and a fish": the "friendship" of diver Jean Delmas with a twenty-five-kilo grouper, Jojo, given in English the name Ulysses (ca. 1:13:53).[89] The divers encourage the curiosity of

FIG. 2.23. Jacques-Yves Cousteau and Louis Malle, *The Silent World*, ca. 24:05. Pathos: a blowfish dying in the atmosphere of air.

FIG. 2.24. Jacques-Yves Cousteau and Louis Malle, *The Silent World*, ca. 1:14:35. Jojo, a grouper, belongs to a species that Cousteau calls the "intellectuals" of the sea (Marden, "Camera under the Sea," 165).

Jojo by feeding him until "Ulysses grew to be a nuisance. It was hard even to make a picture of him," because he would bump against the camera, according to Marden, who was along on the *Calypso* for the sequence with Jojo/Ulysses, which was shot on a stunning coral reef near "Assumption Island, 240 miles from the northern tip of Madagascar." Marden recounts that "Cousteau called the big groupers the intellectuals of the sea," and the narrator speaks of Jojo with affection.[90] Eventually, the divers playfully put him briefly in the shark cage so that they can feed other fish as well. "Jojo was at home with us. We were sorry to leave him" (ca. 1:19:58).[91]

Multiple images show the diver and the grouper in profile, as if exchanging glances. We can look over Delmas's shoulder at Jojo and watch their seemingly coordinated "dance" around the meat sack that cements the relationship. When Jojo and Delmas are first responding to each other around the sack, the camera catches the fish turning in front of it, enabling us to see how the grouper flicks its eyes. For a brief moment, we look at the fish head-on, but the film does not integrate this glimpse into a montage, pairing Jojo's eyes with those of Delmas in his mask, and thereby suggesting emotional exchange between the two (fig. 2.24). *The Silent World*'s innovative fish point of view may mimic what marine creatures see when they swim, but the film nonetheless stops short of suggesting the divers have access to their perspective, in an existential sense.

2.4. *Diving to Adventure*: Romanticism beneath the Waves

If you walked into the main building of the BBC in London in 1956, you might have seen a striking poster, designed by Muriel Madden, the wife of senior BBC writer, producer, and programmer Cecil Madden (fig. 2.25). At the center of its crowd of TV personalities is a shark, framed by Hans and Charlotte "Lotte" Hass. All three star in *Diving to Adventure*, in what one reviewer when it aired called "the most exciting series of submarine films ever screened."[92] In choosing *Diving to Adventure* to highlight in the poster, Madden was selecting from an extraordinary year for nature documentary on TV, with several celebrated programs, in which, as critics noted, "[w]ild life provides fascination and thrills."[93]

Hans Hass, who appeared in the series with his wife and fellow dive pioneer, Lotte, was an innovator in diving equipment, photography, and underwater filming starting in the late 1930s and throughout the 1940s and 1950s. Although he is not well known in the English-speaking world today, Hass had been making underwater films for the public since his short 1942 *Stalking Under Water* (*Pirsch Unter Wasser*) and his full-length *Humans among Sharks* (1947/48). Hass was the first biologist to study and film sharks extensively in the wild and worked without the protection of a cage or other barrier. In 1951, his full-length *Abenteuer im Roten Meer* (appearing a year later in English with the title *Under the Red Sea*) "won first prize at the Venice Film Festival for Best Documentary."[94] His *Unternehmen Xarifa* (1954) (appearing the same year in English as *Under the Caribbean*), was one of the first feature-length underwater color documentaries, and, like Folco Quilici's *Sixth Continent* (*Sesto continente*), predated Cousteau's *The Silent World* by two years.

Diving to Adventure was the first miniseries about the underwater realm.[95] The "hugely popular" shows brought the third dimension of the seas into the homes of millions of viewers on Friday evenings at 9 p.m. in April–May 1956.[96] An internal BBC survey of audiences after the first episode estimated that "the audience for this broadcast was 20% of the adult population of the United Kingdom, equivalent to 50% of the adult TV public," and that the program met with an enthusiastic reception, comparable to "the figures for . . . Captain

FIG. 2.25. Muriel Madden, "TELEVISION 1956 BBC Montage." Photograph © Hans Hass Archive HIST. In this collage, the stars of 1956 are a shark framed by Hans and Lotte Hass.

Jacques Cousteau's illustrated talks (broadcast during the winter of 1954/55) on his recent expedition to the Indian Ocean."[97] Torma comments that "underwater film[s] transformed the oceans into a visual commodity," and television offered its consumption in domestic ease.[98] British marine biologist Trevor Norton reminisces: "Television was to change my life for ever with a series of films in which Hans Hass went *Diving to Adventure*. Each week he swam with giant manta rays and moray eels, and his luscious wife, Lotte, perched on mounds of corals while sharks circled ominously. . . . I coveted Hass's beautiful yacht and the freedom to rummage below on tropical reefs. And then, of course, there was Lotte."[99] Adult reviewers shared his enthusiasm: "Exotic creatures in the Red Sea, old wrecks and the surge of the white yacht rouse the adventurer in all of us."[100]

Cousteau and Hass were both creative innovators in dive and camera technology of the era. Yet the imaginative access to the undersea offered by Hass's spectacle was quite different from that of Cousteau, as the reviews quoted above suggest. Watching *The Silent World*, spectators felt themselves, in the words of the *New York Times* critic Bosley Crowther, a "breathless companion of the modern mariners" on the frontier of deep-sea exploration.[101] Hass's submarine camera and dive technologies were as influential in this era as those developed by Cousteau (although the rebreather was more dangerous to use and limited to specialized diving, in contrast to Cousteau's compressed air tanks, widely adopted by general publics). At the same time, *Diving to Adventure*'s appeal downplayed the technical feats of its creator. Rather, it showcased attractive young divers in an awe-inspiring paradise, drawing on romantic tropes of landscape and the love of nature to celebrate new undersea access along with a new dive lifestyle, open to women as well as men.

Each episode of *Diving to Adventure* featured studio introductions intercut with on-location sequences. In the studio, the Hasses might exhibit coral specimens and show the equipment they used such as cameras, compressed air cylinders and a harpoon. They also emphasized the accessibility of diving by demonstrating a snorkel, goggles, and flippers. "Hass films send viewers out to buy flippers," to quote a headline in the *Oxford Mail*, May 4, 1958, publicizing the episode aired that night. The dangers of such enthusiasm appear in a May 8, 1956, article in the *Yorkshire Post* reporting that the town chairman of the Water and Baths Committee of Brighouse Town Council "gave a warning to children not to attempt underwater swimming," at a council meeting.[102] In issuing such a caution, the council cited "TV films, which have been shown recently [and] will probably encourage a number of boys and possibly girls, to believe that they can become expert at underwater swimming."[103]

Unfortunately, the six episodes of *Diving to Adventure* are not readily available today. I have been able to screen five, thanks to the generosity of the Hans Hass Institut für Submarine Forschung und Tauchtechnik and the Historical Diving Society, which hold copies in their archives.[104] Footage for the on-location segments of the series is largely taken from Hass's films, which are more available and which will give the reader reference points—although the film footage is edited and narrated differently in the more polished BBC production.

In the third episode, which aired on April 20, 1956, about Hass's first trip to the Red Sea, we find the sharks and manta ray so memorable that they were recalled many years later by Trevor Norton, cited above. The on-location sequences of this episode are constructed out of footage from *Under the Red Sea*. The producers reused even material that was easier to obtain than undersea views, such as imagery of Port Sudan, the jumping off point for the expedition. The Hasses were among the first to promote dive tourism to support their research, and the episode also uses some footage from a diving trip that the Hasses organized and guided in 1955. A review in the *Manchester Evening News* noted that Hans and Lotte "lead Red Sea 'safaris' for tourists with conducted tours of shark land at £400 a time, including air fares," although the 1955 expedition was the only one they were able to launch. In contrast to Cousteau's navy divers, the safari included wealthy tourists and prominent figures in the world of photography and film, such as Swiss photojournalist Yves Debraine, the great Danish filmmaker Carl Theodor Dreyer, and movie actor Paul Christian (Paul Hubschmid), who had just starred in Italian director Romolo Marcellini's *Il tesoro di Rommel* (1955), where Hass did underwater camerawork.[105] "I know many of you will be fascinated by Dr. Hans Hass and Lotte's daring exploits under the Seven Seas and will be tempted to follow their example," observed N. Oscar Gugen, the founder of the first British Sub-Aqua Club (which counted thousands of members by that time), in a forty-five-second trailer for the series. Gugen connected the underwater environment with Britain's historical dominance in maritime empire. "Although our climate and temperature of the water may not be as suitable at all times as the seas of other countries," Gugen exhorted Britons, too, to "exploit" the undersea "in accordance with a longstanding tradition" whereby "we should take a ruling hand under the Waves."[106]

FIG. 2.26. Hans and Lotte Hass on the studio set of the BBC's *Diving to Adventure*. Photograph © Hans Hass Archive HIST.

Carefree Adventurers

The great fascination of *Diving to Adventure*, praised by multiple reviewers, was the exhibition of underwater sights never seen before, let alone filmed: "the lure of the unknown."[107] Viewers were also captivated by the explorers, starting with Hans and Lotte Hass. On the BBC living room set, as in publicity photos, the couple appear as a mild-mannered professor and his gracious, demure wife (fig. 2.26). Once on—and in—the water, however, the Hasses became "audacious natators," according to Crowther's review of *Under the Red Sea* in 1952. Crowther noted that Hass had the "good theatrical sense to fill his picture with as much dramatic and amusing personal detail as pseudo-scientific demonstration of the weird and tingling wonders beneath the sea."[108] These appealing characters, who also appear in *Diving to Adventure*, included "two frisky young fellows [Leo and Jerry] and a beautiful and buoyant young blonde," as well as Hass himself, "a bearded young man with stringy, disordered hair, [who] circulates in a sort of gleeful frenzy."[109] The dangers faced by the divers are real, and yet Hass's "audacious natators" seem as much at play as at work. They are shown, for example, lounging on deck in their casual beach attire, although if such sequences were framed differently, they might have emphasized that the divers were recovering from strenuous activity. Underwater, Hass's films intersperse sequences of exploration and danger with beautiful, smooth swimming; playful dancing; and practical jokes, which are taken up in the TV series. Such human interest is well-suited to the intimate scale of TV. From the 1930s, the BBC had organized programs using "human interest and rapid tempo," and it seems as if Hass was already working with a TV aesthetic in his films.[110]

Marilyn Monroe of the Deep Sea

Lotte Hass's role as central participant in diving adventures was an innovation in the history of submarine exploration. As Torma observes, "[H]er underwater presence challenged the male-centric cinematic representation of diving and filming."[111] She was, moreover, the first in a line of adventurous female divers in narrative films, from *Underwater!*, starring Jane Russell, to the successful Bond franchise, such as *Thunderball*, to Jacqueline Bisset in *The Deep* (1977) and beyond.[112]

Women's appearance in films about diving owes a great deal to the commercial potential of the bathing beauty. Esther Williams, an Olympic hopeful who had turned to acting when the 1940 Olympics were canceled because of World War II, had popularized water ballet and synchronized swimming, and went on to film stardom.[113] According to Lotte Hass's obituary in the *New York Times*, "When she was about 19, she answered an advertisement from Mr. Hass, who was seeking a secretary. After he hired her, she implored him to let her come along on his underwater expeditions. He demurred—an oceanographic vessel, he said, was no fit place for a woman. In the end, after learning to dive, she wore him down, assisted by his film company, which quickly saw the value of a beautiful, submersible star."[114]

Hass's producers had calculated correctly. The contemporary British magazine *Picturegoer* called Lotte Hass "the Marilyn Monroe of the Deep Sea," in an article announcing the making of *Under the Caribbean*. This article, like much of the coverage of the time, credited Hans Hass as "underwater film-man" and organizer of the adventure, while "Lotte will star against backgrounds of whales, sharks, and octopuses."[115] A poster for *Under the Caribbean* foregrounds a cheesecake image of Lotte in a bathing suit (fig. 2.27). At the same time, connoisseur audiences accepted Lotte's importance as a diver on the expedition. When the *Skin Diver*, the first US magazine devoted to scuba diving, founded in 1951, reviewed *Under the Red Sea* in its January 1953 issue, it featured Lotte

FIG. 2.27. Publicity poster for the English version of *Under the Caribbean*. Photograph © Hans Hass Archive HIST.

in the first underwater photograph of a woman scuba diving to appear on the cover of the magazine (fig. 2.28).

Lotte's presence on Hans's team, moreover, was not an exception that proved a rule. In contrast to helmet diving, where women were absent until the 1970s, and in contrast to the space program, that other cosmic frontier in the same years, the scuba revolution featured women's participation from its early history.[116] While helmet divers needed sheer strength to support their heavy gear, strong swimmers excelled at scuba diving. Photogenic athlete Zale Parry was featured on the cover of *Sports Illustrated* in 1955 after reaching a record depth for women in 1954 (209 feet). Parry helped pioneer the first hyperbaric chamber and went on in the later 1950s to do stunt diving for

ℭ𝔥𝔢 SKIN DIVER

A MAGAZINE FOR
SKIN DIVERS AND SPEARFISHERMEN

Vol. 2 No. 1   JANUARY, 1953   25c

Read and Written by the World's Greatest Sportsmen

FIG. 2.28. *The Skin Diver: A Magazine for Skin Divers and Spear-fishermen* 2, no.1, (January 1953). The first cover of this magazine (and possibly any magazine?) featuring a woman scuba diving: Lotte Hass.

underwater movies. In *Sea Hunt,* she both acted and was the stunt diver for the many female characters who accompanied Mike Nelson underwater throughout the four-year series. Marine biology was one of the areas of science that was the most accessible to women throughout the twentieth century, and women took up the new technology of scuba to further their research. While the most famous female marine biologist of the era, Rachel Carson, did not dive, shark biologist Eugenie Clark published a best-selling memoir including her research undersea, *Lady with a Spear* (1953). By the middle of the 1950s, moreover, women started to take up diving and underwater photography, ac-tivities at the adventurous end of beach-culture lifestyle. Dottie Frazier was an early female member of the Southern California dive milieu, a commercial fisherwoman, scuba instructor, and free diver.

Despite women's presence on the undersea frontier, they maintained a subordinate role throughout. As Torma notes, Hans "is still the one who explains the technology and scientific details" to Lotte in their films, even as she "dives and takes picture of the ocean."[117] Richard Crosby's review in the *Skin Diver* of *Under the Red Sea* is tellingly condescending. It praises at length the film as a "mile-stone in the art of skin diving" and "pictorial proof of the personal courage of" Hass, with a paragraph on the "serious and charming skin diver," who can answer all the "advanced diving physics" questions Crosby throws at her, and thus proves to be "more than a cheesecake model on the screen."[118] Crosby's condescension reflects not just the stereotypes of the time but also how Hass's films, like *Diving to Adventure*, cast Lotte in subordinate roles that would recur throughout exhibition of women's underwater prowess in the 1950s. These roles include a willing student who could enable the explanation of technologies, an admirer of undersea beauty, and a headstrong young woman who would have to be rescued from her own imprudence. If Norton recalled Lotte perched on the coral from his childhood crush, a female scuba diver appeared in such a pose on the rusty anchor of ship wreckage featured on the cover of the *Skin Diver* in July 1956, less than three months after *Diving to Adventure*. For some viewers perhaps (in an eroticism that the Bond franchise would develop) scuba technology took on a fetishistic appeal. A headshot photo of Lotte accompanies Crowther's review of *Under the Red Sea*, where she wears as a necklace a double hosed regulator.

Submerging Romantic Landscape

Hass's sequences exhibiting the body freed to enjoy the sea continue a Romantic lineage celebrating human harmony with nature. In *The Lure of the Sea* (English translation, 1994), Alain Corbin traces the emergence of the beach and bathing as a new practice of hygiene and sociability in the Enlightenment and Romantic eras. Hass's exhibition of hitherto unseen undersea wonders in his films, as in *Diving to Adventure*, drew on the organizing trio of landscape aesthetics in the Enlightenment/Romantic era: the picturesque, the beautiful, and the sublime.

In the late eighteenth and early nineteenth centuries, the picturesque aesthetic, which highlighted variety and encouraged curiosity, inspired travel and tourism. In the pithy summary of the University of Texas Austin's Blanton Museum, a "picturesque view contains a variety of elements, curious details, and interesting textures, conveyed in a palette of dark to light that brings these details to life. In the mid-eighteenth century, tourists who followed the cult of the picturesque traveled to untamed areas of the British Isles in pursuit of this visual ideal."[119] Hass recognized, whether intuitively or consciously, that viewers would relate to this aesthetic amidst the extraordinary, fascinating variety of life found on reefs. Further, he recognized that northern European audiences already sought out such an aesthetic in their tourism of the global south, whose coasts were his submarine gateways. Thus, in *Diving to Adventure*'s episode 3, Hass used the stereotypes of first-world orientalist travelogue to describe Port Sudan: "How picturesque it all was with its brilliant contrasts of light and shade, its cafés and shops, and its jumble of man and animals" (ca. 4:50).

When the episode, following *Under the Red Sea*, then took audiences to view the coral reefs, Hass devised an arresting submarine picturesque using montage, since the shallow depth of field where imagery holds clarity could not contain those contrasts in a single frame. In episode 3, this montage starts around 7:46, following a shot in which the shimmery view of a shallow reef visible topside dissolves into an exhibition of the marvelous variety visible with the underwater camera. The coral species are slender and bulbous, reticulated and smooth, pointed like antlers and branching like tree, with a range of textures that stretch the limits of everyday terrestrial vocabulary (fig. 2.29). Some corals are still, others move gently or sway, and the chiaroscuro of light as it comes down from the surface dapples the scene, gently oscillating through the moving water.

Hass's picturesque employs a piquant and entertaining tone, in contrast to the surrealist vision of corals as an emblem of convulsive beauty discussed in part 1.[120] Rather than posing trouble for Enlightenment categories, coral for Hass offered the chance for a lesson in marine biology. In the studio sequences of episode 3 (starting ca. 2:26), Hans and Lotte describe the physical makeup of coral and show varieties in skeletons, contrasting massive atolls to corals, which like their name, look like delicate lace; some corals growing in "big brown sheets like the acorn coral" (ca. 3:48); organ pipe coral, "which stands up in straight tubular clusters" (ca. 3:58); and fire coral, "which looks underwater bright and yellow" (ca. 4:03). Once the camera descends beneath the surface, we can better understand what we see. Lotte's voiceover infuses knowledge with marvel when she invites us along on a tour of "the underwater wonderland which Hans has so often told me of" (ca. 9:15). "I had seen it all before in Hans's pictures of course, but how much more wonderful it was in reality" (ca. 9:42).

FIG. 2.29. Hans Hass, *Under the Red Sea*, ca. 15:56. Underwater picturesque: varieties of coral.

Along with the tour of picturesque marvels, audiences were captivated by depictions of diving as sublime. For Enlightenment and Romantic aesthetics, the sublime was the pleasure taken in representations of overwhelming power, to the point of terror. Hass was versed in the Romantic sublime and its use to portray the fatal depths. Indeed, in *The Undersea Adventure* (1951), Cousteau's fellow diver Philippe Diolé relates, "Hans Hass tells how, in the Caribbean sea, at a moment of great pain, he recited Schiller's *Diver* to himself over and over again."[121] In this ballad, Friedrich Schiller recounts the fatal lure of the depths for an overly daring young man.

One important visual expression of the sublime for Romantic artists was plunging verticality. Thus, Caspar David Friedrich's *Chalk Cliffs on Rügen* (1818) portrays the vertigo of spectators looking down from the heights of cliffs that fall off into the sea. Similarly, J.M.W. Turner's images of the ravines in the St. Gotthard Pass emphasize the terrifying chasms and the danger of falling into the abyss. When Cousteau and Malle show the third dimension of the sea in their opening to *The Silent World*, they associate it with the emancipation from gravity offered by flight. Hass, in contrast, uses it to draw a diagonal line of descent into an extreme environment testing human capacity and limits.

Hass had used such a line of descent from the beginning of this career, as exemplified by the photo on the dust jacket of his 1939 book, *Hunt Underwater with Harpoon and Camera* (*Jagd unter Wasser mit Harpune und Kamera*). The environment surrounding Hass, the diver in this image, could be considered a version of the submarine picturesque, incorporating rough and smooth, light and dark, fish and reef—in short, "a variety of curious details, and interesting textures." And yet, Hass's body creates a diagonal line accentuated by his harpoon. This line bisects the image, separating humans and the book's title from the natural environment, and showing the diver on a bold descent into the unknown.

Hass would intensify the angle of descent along with the danger in episode 3 of *Diving to Adventure*, when he undertakes, in his words, "a really deep dive see how far down one could go without breathing apparatus. Pressure of course is the great enemy, on the ears, on the eyes, on the lungs, and since the delicate organ of balance is inside the ear, one scarcely knows whether one is on one's head or one's heels" (ca. 12:17). The sequence frames the beginning of the dive against the edge of a reef that accentuates the almost 90-degree angle of his plunge (fig. 2.30). Reviewers noted the impact of the sequence. In the words of the *Birmingham Mail*: "Although Hans and Lotte Hass are actually present to introduce . . . [the films], one still wonders if they are going to survive their adventures." "Two instances last night must have been watched with bated breaths in many homes"; the first, "the long, long dive without apparatus."[122]

"To make anything very terrible, obscurity seems in general to be necessary," the influential Edmund Burke theorized in *A Philosophical Enquiry into the Origin of Our Ideas of the Sublime and Beautiful* (1757). "When we know the full extent of any danger, when we can accustom our eyes to it, a great deal of the apprehension vanishes."[123] After episode 3's shot of Hass plunging along the sheer 90-degree cliff of the reef, the camera shows him disappearing from view, into darkness below. The visual imagery incorporates blackness and a fog of varying shades of gray that admit perhaps a glimpse of the diver's body. Spots further confuse the view, as if from lights or bubbles. The narration of the episode links its visual obscurity to the disorientation of the mind in such a free dive to the limits: "fantasies fill the brain, and always at the back of one's mind is the realization that one has the return journey to make upwards" (ca. 12:50). On the yet more terrifying ascent, Hass represents vertigo in a more radical way than that of Romantic imagery, turning the sublime to motion sickness. Showing Hass's point of view in the final push upward to the surface,

FIG. 2.30. Hans Hass, *Diving to Adventure*, episode 3, ca. 12:17. Underwater sublime: "a really deep dive."

the camera rotates vertiginously 360 degrees, as it focuses on underwater views of the boat's hull and engine. This rotation, using the mobility of the movie camera underwater that delighted Malle, expresses the "last memory" of Hass, near if not blacking out "as I gradually neared the surface" (ca. 12:58).

The beautiful is the third aesthetic in the Romantic trio for showing landscape. Hass calls on it to exhibit what critic Reginald Pound appreciated as "the beauty of form and movement," a distinctive aspect of underwater existence that the movie camera is suited to reveal.[124] Hass's films focus repeatedly on the grace both of the divers and of the underwater creatures. "Deep sea 'ballet' made good viewing"[125] is a headline from a review of the fourth episode, while a review of the fifth episode begins with "Underwater Grace."[126] While critical of the "clumsy" studio scenes, the reviewer praised "the underwater gamboling of these two modern explorers," declaring, "[u]nderwater swimming is a graceful spectacle with an abundance of lightness."[127]

Reviewers most often highlighted Lotte as the epitome of such movement. They referred to her repeatedly as a "mermaid," and a preview by journalist Alan Morris introduces her to the public as "more at home in a swimsuit than

Esther Williams."[128] The series included portions of *Under the Red Sea*, leading Crowther to remark that "[t]he varieties of fish that are shown have a lively pictorial fascination . . . [b]ut the figure of Miss Baierl, floating smoothly along the crags of coral in the underwater world, makes an equally fascinating and dramatic contrast to the life that is there."[129] Although Lotte commanded more critical ink, the male divers on Hass's undersea expedition are just as accomplished. The film shows off their exhibitionistic athleticism (one indeed wears leopard-print swim trunks). Hearkening back to the cinema of attractions, Leo and Jerry fill the screen with acrobatic maneuvers underwater such as a somersaults and back flips, conveying their pleasurable freedom moving in three dimensions. Accompanied by upbeat popular music, they perform an energetic and slightly comical pas de deux.[130]

Gliding comes naturally to marine life, perhaps, but Hass's camerawork emphasizes the animals' movement as a dance, from fish to the most extraordinary creatures of the sea, such as the giant manta ray and shark. While in *The Silent World* a jerky, handheld camera captures sharks in the midst of feeding, for Hass, sharks epitomize majesty. Lotte marvels at a blue shark, whom she calls the "king of the fish" and whom Hans deems "a very graceful and handsome creature" (ca. 15:58, 16:12). Footage of sharks swimming exhibits their power and authority in what one reviewer called "the slow, silent procession of sharks [passing] . . . their almost defenceless watchers" (indeed, we see Hass's alert divers poised with harpoons in this sequence).[131] Without a sense of the threat posed by sharks, there would be little to say about the fluid simplicity of their presentation in Hass's footage, which inaugurates environmental awareness about the shark's importance in the marine ecosystem, along with a recognition that humans are not their preferred prey. As Hass says in the studio section of this episode, "I formed a theory quite early in my student days that this monster might not be so black as it had been painted. Few wild creatures are. [They] . . . would much rather leave you alone, provided you leave them alone, and they only fear starvation" (ca. 17:00).

Hass is so compelled by dance underwater that he even creates a Viennese gala by placing microphones on the seafloor to see if fish respond to human sounds beneath the sea. While indifferent to many topside noises, the fish all move as if they were synchronized swimmers when the expedition tries out Strauss's water waltz, "The Blue Danube." When Crosby reviewed *Under the Red Sea* for the *Skin Diver*, writing for audiences knowledgeable about the submarine environment, he remarked on this sequence's extraordinary impression. He declared, "As the hundreds of these musically-stirred listeners closed in around the diver, swimming in a circle around him like a tall, revolving curtain, it made the most beautiful spectacle I had ever seen. These unrehearsed, wild creatures of the deep actually seemed to dance in graceful calm around their stranger!"[132]

2.5. *Sea Hunt*: TV's Watery Epic

Spectacles considered to this point in part 2 have used the new capabilities of underwater film to reveal hitherto unseen environments, inhabited by both strange and familiar marine life, and have found innovative ways to imbue these environments with a variety of moods. The works have also peopled these settings with new characters, like Hass's carefree adventurers and Cousteau's professional team. With new dive access and new possibilities for the underwater camera, these works devise innovative organizations of settings and new shot patterns that adapt and transform previous submarine fantasies.

Along with characters, setting, and shots, narrative is an important element of film, the seventh art that unfolds in time. Those involved in making films throughout the 1950s were experimenting with how to arrange narratives underwater and undersea. Science fiction was one way to imagine this new environment first introduced in *20,000 Leagues*, even if elements of Verne's fiction, such as diving untethered and an electrical submarine, had by now become fact. Filmmakers would try out other science fiction possibilities, whether it was the discovery of primordial creatures in the Gill-man trilogy launched by the *Creature from the Black Lagoon* (1954) or post-nuclear fantasies suited to the atomic age, such as Roger Corman's *Attack of the Crab Monsters* (1957). The episodic organization of seafaring adventures from the global age of sail shaped *The Silent World* and *Diving to Adventure*. Such narratives were unified by what mariners in their logs called "remarkable occurrences," designating various unexpected dangers and new sightings encountered in regions of the globe hitherto uncharted by Western societies.

The production in this innovative decade that would devise a way of narrating submarine adventure so robust that it could be reproduced and varied turned out to be a TV series rather than a film. When *Sea Hunt* was first proposed in 1957, networks were skeptical that diving fiction could hold viewers' attention. In the words of producer Ivan Tors, "the idea seemed to be all wet." The networks liked "the first story, but they didn't think we could keep a show going for very long under water."[133] However, Ziv Television Productions' *Sea*

Hunt ran for over four years, 1958–61, with 155 episodes. In 1959, *TV Guide* called it "[o]ne of the most popular syndicated shows in the history of television."[134] Tors told the periodical, "We weren't sure at first how much underwater stuff we could get by with. But we soon found out that that was what the audience wanted—water, water, and more water."[135] The series made Lloyd Bridges's career, and its fame endured. Three years after the last episode, the Transportation and Travel Pavilion of the New York World's Fair of 1964 advertised as one of its "highlights," "UNDER THE SEA. In a huge tank, skin divers put on an underwater drama. Lloyd Bridges, star of the television show 'Sea Hunt,' narrates the show on tape, and from time to time appears in person."[136] According to the informative although now defunct website of popular scuba historian Bill Jones, *Sea Hunt* was "the country's most successful first-run syndicated TV show until *Baywatch* (1989–2001)."[137]

The impact of this series was so great that it is arguably the post–World War II moving picture equivalent to *Twenty Thousand Leagues under the Seas*, the master fiction popularizing the revolution in underwater access from the aquarium era. Just as *Twenty Thousand Leagues* drew up a chart of the modern undersea imagination that would influence generations of readers for the next century, so *Sea Hunt* created a model of undersea adventure that would influence narrative film and TV to the present day. While *Sea Hunt* has largely been ignored in scholarship, the series is worth the attention of cultural historians for its ideological construction of undersea exploration, work, and leisure. In keeping with my focus on underwater aesthetics, I want simply to explain the series' invention of an enthralling and repeatable narrative structure enabling it to generate episode after episode. The key to this structure was in the teaser (with a few variations) spoken at the end of each episode by Lloyd Bridges: *Sea Hunt* offered "action-packed stor[ies] . . . of underwater adventure."[138] *Sea Hunt* submerged the adventure patterns developed across a long history of sea fiction in the perils and excitements of submarine conditions and used as their content the activities now possible in the depths with the new access of the postwar era.

The basic structure of adventure narrative is a chain of problems and solutions that pose new problems in their turn—in a process that vindicates the ethos of its protagonist, even as it engages the reader in problem solving alongside the characters in his or her imagination.[139] From Defoe's *Robinson Crusoe* (1719) to the Patrick O'Brian Aubrey-Maturin series penned at the end of the twentieth century, transatlantic writers generated problems and solutions from the realities of sea's extreme environments and the technologies used to navigate them—including unknown waters, coasts, and their inhabitants; storms; wrecks; naval skirmishes; and creatures from whales to sharks to giant octopuses. Sea fiction writers also found problems and solutions in what Joseph Conrad called "the human element" of sea travel, fraught with murky personalities, whose motives often conflict with professionalism and safety.[140] *Sea Hunt* absorbed this pattern from centuries of overseas navigation narrative and transferred it to the underwater realm. Mike Nelson exhibited the practical capability I have called "craft"—which has been vindicated by the heroes in sea adventure fiction from Robinson Crusoe to James Fenimore Cooper's John Paul Jones and Robert Louis Stevenson's Jim Hawkins—in contending with problems in areas ranging from oceanography and marine biology to military operations, commercial exploration of the depths, leisure diving, and even underwater movie making.

Novels and tales use narration, while TV and film prize showing over telling. Mike Nelson's voiceover set up each episode, but action became the favorite means of showing in an environment where characters were submerged in the climactic scenes and hence rarely able to speak. Above all, film and TV drastically reduced the complexity and scale of adventures in novels. While sea fiction played out problems and solutions over the many pages of a novel—Robinson Crusoe, at the start of the tradition of adventure in extreme environments, spends twenty-eight years on his island—*Sea Hunt* compressed its dramas into twenty-five-minute episodes.

A hectic production schedule further discouraged intricacy. A *Sea Hunt* episode was shot in three days, and the first three seasons had thirty-nine episodes each, while the last had thirty-eight. By April 1959, an article in *TV Guide* states, sixty-five episodes had already been filmed. The legendary waterman and underwater photographer Lamar Boren did most of the camerawork.[141] To facilitate rapid shooting, Boren developed "a balanced, watertight camera apparatus which can be utilized at all depths and pressures and which is small enough to be hand-held by one man. Without it, most of the underwater shooting could not be accomplished."[142] Lighting too was simplified, even from the lighting used in *The Silent World*, with natural light used where possible.[143] The compression also necessitated minimal backstory or context.

TV Guide pointed to what it called the "semidocumentary" quality of the series: "Producer Tors has wisely chosen to keep proceedings on a semidocumentary level, packing each episode with genuine ocean lore and going in heavily for the underwater stuff."[144] As I have mentioned, the show was not at all documentary when it came to the setting, which did not seem to bother its viewers. However, publics were less accepting of violations of diving protocols, evincing their interest in scuba. In Tors's words, "What we do get on almost every episode are letters from skin divers all with the same complaint: Why do we let Mike Nelson dive alone? I know it's against skin-diving safety rules"; however, Tors admitted, if "Mike Nelson has a companion to get him out of trouble—poof!—there goes the plot."[145] The "[n]eed for a writer to be well-researched in the technicalities of such series," to quote an article from *Variety*, explained the reason "for settling upon one major scripter," Art Arthur.[146] Markups of scripts reveal that someone combed through them with an eye to accuracy. For example, written comments on a script that became the last episode (39) of season 2, "The Raft," correct or approve the symptoms of oxygen deprivation in Nelson's free ascent. "No," is written in red pen next to the moment when Nelson releases "bubbles to ease lung pressure" as he is starting to run out of air (presumably since there is no need to release air on a free dive ascent, in contrast to scuba, where the diver needs to breathe continuously). An asterisk confirms the accuracy of the spots in Mike's vision as he starts to black out in the ascent, symptoms that Hass had included in his sequence on the deep free dive in episode 3 of *Diving to Adventure*.[147]

Sixty Feet Down: The Poetics of "Not Enough Air"

One source of reader interest in sea fiction is the way it dramatizes unfamiliar technical knowledge, notably from the work of seafaring, as well as knowledge of unknown environments, peoples, and creatures that Western mariners encountered in their travels.[148] *Sea Hunt* similarly shares technical knowledge about diving and the new activities practiced beneath the surface of the sea, which provide intriguing context as well as problems that needed to be solved. These problems anchor many of the "action-packed stor[ies] of underwater adventure" that prove Nelson's heroism (ca. 24:47). *Sea Hunt* adds suspense to this pattern of sea fiction specific to the environment where its drama plays out. Because the atmosphere of water is toxic to humans, the need to solve problems takes on an urgency suited to the twenty-five-minute episode frame, as soon as a character is submerged with barriers to surfacing. *Sea Hunt* utilizes this potent blend of adventure problem solving and environmentally generated suspense from the very first episode of the first season, "Sixty Feet Below."

In "Sixty Feet Below," the event holding the characters underwater is the crash of a valuable military plane that malfunctions, the "XF-190" (modeled on a notorious test plane of the mid 1950s).[149] The plane is first glimpsed by

Mike Nelson on his routine day job at Marineland of the Pacific. Sighting the plane overhead, Nelson remarks, "it's kind of funny how the jet pilot's problem of survival is same as mine. Oxygen, taken for granted at sea level, but of vital concern at 40,000 feet up . . . or 40 feet down" (ca. 1:31). When the plane crashes into the sea, the military brings Nelson in by helicopter to save the valuable ten-million-dollar craft, and we see the parallels played out. Nelson finds the pilot, one Jim Cook, still alive, trapped in the plane and breathing from the oxygen supply used by test pilots for high-altitude flying. Nelson must figure out how to release Cook from the aircraft, whose canopy is weighted down by the intense pressure of the depths, before Cook's oxygen runs out. In these efforts, Nelson is thwarted by the impact of the aquatic at-mosphere on topside tools. He first has the idea to bring the plane to the sur-face with one of the winches on the boat. However, once logged with water, the plane exceeds the capacity of the winch, and the gear breaks. Then Nelson proposes opening the canopy. First he tries bashing the canopy with an axe, but he cannot achieve enough force underwater. Finally, Nelson takes down a blowtorch to cut the pilot out. In the nick of time, the pilot points to free oxy-gen in the canopy, which the blowtorch would ignite. The problem is solved when a control boat member realizes that Nelson can use the blowtorch to create a hole in the part of the cockpit that the water has now entered, where Nelson will find the release lever. As the pilot is floundering and gasping, Nel-son succeeds in opening the canopy. Nelson swiftly gives Cook the regulator of a spare lung, and the control boat crew hauls the pilot to the surface.

While the literature of sea adventure revels in the granularity of complex challenges that can be drawn out over many pages, the tools used here by Nelson and the topside crew are rudimentary. A cable, an axe, and a blowtorch would scarcely be the elements of gripping adventure on land in 1958—but they become dramatic when they behave so differently in the depths. The situ-ation is also dramatic because the spectator is aware that each rescue attempt absorbs valuable time. "And time—the minutes were running out," Nelson declares, in the course of efforts to save the pilot (ca. 13:10). The limited air supply pressing the episode toward conclusion holds the spectators in thrall. French narratologist Roland Barthes analyzes suspense as the interplay of for-ward motion and blockage, creating what Barthes characterizes as "paradoxi-cal" temporal tension.[150] Barthes dissects a novella by Balzac, *Sarrasine*, that slowly unravels the mystery of the central character's identity. Barthes shows how the narrator, prompted by desire and curiosity, presses the narrative for-ward in his hunger for information about a glamorous opera singer who attracts him (a castrato, it will be disclosed at the story's end), while what appear to be clues in fact conceal the truth even as they seem to advance its disclosure.[151] In "Sixty Feet Below," the dwindling supply of air drives the narrative forward, and the blockage occurs with attempts at rescue that fail.

To enhance the sense of drama, "Sixty Feet Below" includes reminders of limited time throughout the episode. These cues correspond to a type of narrative figure that media theorist Hans Wulff terms a "cataphora." Cataphora is a rhetorical technique in which the speaker uses a word that derives its meaning from a word yet to come, such as using a pronoun before giving the noun. In Wulff's application of this concept to moving pictures, cataphora are cues that heighten viewers' anticipation about where a plot is heading, notably plots of danger. Wulff writes, "Dangers are brought into play as possibilities but not as actual developments. The danger is the '*not yet*' of a catastrophe or injury. The activity of anticipation reaches out to this *state of 'not yet'* and tries to give it a more precise definition. . . . What probability do the various alternatives have for the course of events?"[152]

"Sixty Feet Below" uses both verbal and visual cataphora to keep audiences aware of just how much time the pilot has left. As I've mentioned, throughout the series, voiceovers enliven underwater action. After Nelson finds the pilot alive, his voiceover encapsulates, in Wulff's words, "the various alternatives . . . for the course of events": "He would die unless I got him out. Yet only fifteen minutes of oxygen left. But how could he be saved? Could I get him out alive? The odds were impossible but something had to be done. Still a wrong move might kill him instead of save him. The pressure might finish him. The water might drown him" (ca. 12:52).

Visual cataphoras include measurements of time and depth emphasizing the pilot's plight. These measurements start with a close-up of the face of Nelson's depth gauge, which reads sixty feet, when he finds the plane (fig. 2.31). They then include a close-up of a slate on which Nelson writes first the depth of the plane and asks about the amount of air remaining to the pilot; and Cook, after he sees Nelson, making a fist and then spreading his five fingers wide three times, to signal that he has fifteen minutes of air remaining. Topside, we repeatedly see a man in a hat and trench coat nervously looking at his watch, beside the officer on the boat speaking on a telephone with Nelson underwater. As the minutes count down, suspense intensifies with close-ups of the pilot increasingly agitated in his mask, as the water level rises to his mouth.

In this episode, "plot time"—the time it takes to accomplish the actions of the narrative—comes remarkably close to coinciding with "screen time"—the

FIG. 2.31. Felix Feist, dir., *Sea Hunt*, season 1, episode 1, "Sixty Feet Below," ca. 11:30. Mike Nelson's depth gauge: "The pressure might finish him." "And time—the minutes were running out" (ca. 13:00).

time the episode takes to be shown. Nelson learns twelve minutes into the episode that the pilot has fifteen minutes of air left. It takes another ten minutes for him to free Cook. The coincidence of screen time and plot time heightens the suspense for the viewer, because it suffuses every second of viewing with life-or-death potential. The coincidence of screen time and plot time is rare in Hollywood cinema, which cuts from plots "dead time," time that would be part of a drama in life but that does nothing to further viewer interest. The typical divergence between conventional plot time and dead time is highlighted by the few virtuosic exceptions, such as Fred Zinnemann's 1952 *High Noon*, where the clock ticks toward the showdown at the hour that gives the movie its name.[153] However, it is hard to devise plots where all actions are suspenseful.[154] Once humans adventure underwater, dead time is drastically reduced. So, too, is complex action: just a kink in the air hose and dead time springs to life.

Nonetheless, even in the toxic underwater atmosphere, the plot details must be arranged to activate environmental suspense. Thus, compare the cataphora of "Sixty Feet Below" with the scene of danger already studied in *The Silent World* (1956), in which Albert Falco saves André Laban from the rapture of the deep. This episode begins with Laban surfacing, looking unwell. He is soon followed by his dive buddy, Bébert. In a voiceover with a flashback set underwater, Bébert tells the story of how Laban almost drowned. The viewer is

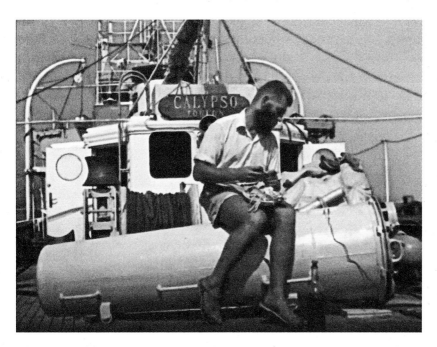

FIG. 2.32. Jacques-Yves Cousteau and Louis Malle, *The Silent World*, ca. 13:14. The ship's doctor feasting on spiny lobsters, "the culprits" of Laban's decompression sickness (ca. 9:27).

enthralled by the film's menacing portrayal of the rockface, dramatized with the conventions of film noir. However, we know Laban survived. Reassured that all ends well, we learn about nitrogen narcosis and decompression sickness, in keeping with Cousteau's interest in teaching audiences about scuba. To further dissipate suspense, the film uses comedy. Decompression tables advise that Laban should spend three hours in a hyperbaric chamber while his companions regale themselves with the lobsters that were "the culprits," in Bébert's words (ca. 9:27). In this sequence, Cousteau and Malle show Laban peering out despondently from the decompression chamber, used as an impromptu picnic bench by the ship's doctor, balancing his plate of lobster (fig. 2.32).

I call the pattern of environmental suspense in *Sea Hunt* "not enough air," in homage to its invention in *Twenty Thousand Leagues under the Seas*. In most episodes, Nemo takes submarine refuge from surface danger. However, in the episode launched by the chapter "Not Enough Air," Verne activates the atmosphere's menace. After Nemo reaches the South Pole, a feat that had not yet been accomplished when Verne wrote the *Nautilus* becomes encased in ice. The *Nautilus* is a proto-nuclear submarine (indeed, the first nuclear submarine launched by the United States in 1954 was named the USS *Nautilus*, although whether it was named after the novel or the Disney film is debated). One resource that this largely self-sustaining submarine cannot generate from the ocean, however, is air. Instead, the submarine has self-contained tanks, which it needs to replenish at the surface. If the *Nautilus* cannot escape, captain, crew, and passengers face the fate of either "being crushed" by the ice, or "asphyxiation."[155]

To enable the *Nautilus* to surface, the resourceful Nemo starts the adventure chain of problem solving, inventing a variety of science fiction responses. Nemo first discharges hot water, which is heated by running water through the coils of the submarine, in order to stop more ice from freezing around the vessel. At the same time, he uses ice mining techniques to chip away at the layers of ice. Eventually, he breaks the *Nautilus* free by filling its water tanks with "100 cubic metres of water," equivalent to "100,000 kilograms" (327). The submarine crashes down through the last layers of thinned ice, and, once beneath it, can make its way around it to the open ocean. In "Not Enough Air," limited oxygen supply underscores the imminence of danger throughout these efforts, which succeed just before it gives out.

In emphasizing the scarcity of time, Verne continues conventions used by adventure narratives in extreme environments to intensify readers' interest— whether the danger is an oncoming storm that can push a ship onto the leeward shore, as in the opening to James Fenimore Cooper's *The Pilot* (1824), or the vicious grip of a giant octopus, called a devil-fish, that threatens to kill the canny fisherman Gilliat who penetrates his lair on a reef in Hugo's *The Toilers of the Sea* (*Les travailleurs de la mer*) (1866). In many narratives, time is a resource that ebbs, as in John Paul Jones's consummate navigation out of a strait called the Devil's Grip at the opening to *The Pilot*. However, narrators of novels have the freedom to suspend the action in dangerous situations in order to pursue other threads. Thus, after the devil-fish attacks Gilliat, inflicting unbearable agony, Hugo stops the action for an entire chapter to discourse on the octopus's menace. He recalls the history of its representation, compares it to other marine monsters and invertebrates, describes its biology, and philosophically transforms it into an expression of horror, terror, and evil. Such a review is entertaining and thought-provoking for the reader, but it leaves the hero expiring in the clutches of destruction. In "Not Enough Air," in contrast, Verne binds the pace of the narration closely to the pace of the environmental danger recounted in the plot.

Measurements of barriers to surfacing, too, are cataphoras in "Not Enough Air," as in "Sixty Feet Down." Depth is one: "The *Nautilus* slowly sank and came to rest on the icy floor at a depth of 350 metres, the depth of the lower surface of ice" (321). The thickness of ice is another: "The following day, 26 March, I continued my miner's work and began on the fifth metre. The sides and ceiling of the ice-cap were getting visibly thicker" (324). However, after hot water starts flowing into the sea to aid the miners, ice measurements suggest hope: "The following day, 27 March, six metres of ice had been torn from the hole. There remained only four metres to go" (326). On the day after that, "[o]nly two metres separated us from the open sea . . ." (327). At the same time, the end of the sentence recalls the unstoppable dwindling of air: "but the tanks were almost empty" (327) and intensifies fear over impending catastrophe.

Like the pilot struggling as the water level rises in "Sixty Feet Below," Verne appeals to readers with the sensations of stress. In "Not Enough Air," these cataphora are narrated rather than shown. As the air degrades, Aronnax describes how "the feeling of anguish in me grew to an overwhelming degree. Yawns dislocated my jaws. My lungs worked fast. . . . A mortal torpor took hold of me" (326). At the final moment, this feeling is suffocation at the brink of death: "My face was purple, my lips blue, my faculties suspended. I could no longer see, I could no longer hear" (328).

Action-Packed Stories of Underwater Adventure

The narrative formula may be simple, but *Sea Hunt* was able to draw from it 155 episodes. In this section, I survey *Sea Hunt* plots, to give a sense of how writers and filmmakers could create so many narratives offering "water, water, and more water" over four years.[156] Dramatically presented thanks to environmental suspense, adventure plot lines explored the widely diverse activities enabled by the new undersea and underwater access at this time, including military, commercial, scientific, criminal, and leisure pursuits. Further enhancing viewer interest, the series offered a new iteration of the cinema of attractions in showing both newly accessible environments, and often just as intriguing, how they were accessed. Fascinating elements of this examination ranged from new technologies to dive and beach clothing. If the series was the most popular TV series until *Baywatch*, it, like *Baywatch*, used water settings to entice with eye candy. Bridges was chosen in part for his physique, and the beautiful Parry, who was introduced underwater uncredited at the beginning of season 1, episode 2, appeared throughout the series. In Parry's words, "I was the gal on topside maybe 12 times on *Sea Hunt* but underwater it was a whole lot easier to have Zale be everybody. The fellows wore wet suits but the girls almost always wore swim suits."[157]

Some episodes hold viewer attention almost purely through the environmental suspense of not enough air, with rudimentary problems involving scuba technology. Thus, season 1, episode 2, "Flooded Mine" takes the tension of running out of oxygen inland. Nelson is summoned to investigate an explosion flooding a mineshaft, where the managers believe all men lost. Instead, Nelson discovers two men alive, trapped in a pocket of air. He uses scuba to reach the miners, but when constraints force him to rescue them individually, he does not have time to obtain the additional tank needed for this unanticipated diving before the oxygen in the pocket runs out. Under duress, Nelson jury-rigs a tank capable of withstanding submergence with materials lying around the mine, including a miner's oxygen tank, which permits him to reach the trapped man. He then brings him to safety using buddy breathing—the sharing of a single scuba tank by two people.

Environmental suspense structures the third episode of season 1 as well, "Rapture of the Deep." Nelson is tasked with hiring workers to mine a fictional underwater mineral, virilium—appropriately named given the macho environment of diving in this era and in the show. In part he is testing their susceptibility to rapture of the deep because of the depths where virilium is found. The climax occurs when one of the recruits succumbs, knocks out Nelson and then tries to kill a rival. Ultimately, Nelson overcomes the intoxicated diver with sheer strength, wraps him in a net, and sends him up to the surface. Along with the dangers of not enough air, environmental drama, if not suspense, occurs when this episode portrays the effect of rapture of the deep.

Sometimes, more complex technologies expand suspense beyond the dangers of the marine environment. Nelson must defuse an undersea navy mine gone astray, first caught on a fishing net, and that then floats away, in "Magnetic Mine" (season 1, episode 21). To the menace of "not enough air" the episode adds the mine's potential to detonate at any moment, another variation on the fundamental suspense protocol of "not enough air," which could be called "not enough time." A similar use of "not enough time" gives interest to "Niko," (season 4, episode 17), in which Nelson must defuse two torpedoes discovered in a West Coast harbor within a twenty-four-hour period, owing to conditions set by Washington. The torpedoes' deadly potential is evident from about 1:32

into the episode, when Mike and a buddy hear one of them ticking. In this variation, there is not a countdown within a known time horizon but rather ambiguity, as the torpedo could detonate at any moment: "We better hurry. With that time fuse we don't know just how long we've got" (ca. 6:17).

Anti-Japanese prejudice was a second theme of "Niko," in an incorporation of topical, social subjects that also marked the series. Yet, as "Niko" illustrates, such themes were generally animated by environmental suspense and/or the exhibition of underwater technologies. Thus, "The Sea Sled" (season 1, episode 5) depicts Hungarian freedom fighters on the run, exploited by underwater smugglers. Nelson helps them escape thanks to the sea sled of the title, a vehicle enhancing underwater mobility. The dangers of the depths then figure in the climax, showing the need for a controlled ascent to avoid "the bends" or "air embolism" (ca. 22:10) after one of the refugees has plunged into extreme depths, cut loose by the smugglers.

The superiority of scuba diving over helmet diving ("hard hat we call it" (ca. 0:38), says Nelson) organizes "Hard Hat" (season 1, episode 14), in which Nelson dives in a small Latin American port for a safe located on a sunken submarine that once carried top-brass Nazi officers. The episode's technical interest derives from Nelson's use of scuba to bring to the surface the historically valuable safe hidden in the recesses of the sunken vessel, out-diving two unscrupulous hard hat divers. "The Sea Has Ears" (season 1, episode 38) explores the expansion of undersea telecommunication and dramatizes the sabotage of an undersea cable linking a chemical plant on an island to the mainland. The layout and vulnerabilities of undersea cables, and the use of sonar to locate the saboteurs, are sufficiently gripping so that the episode does not need environmental suspense. Instead, the climax is an underwater chase scene, concluding in a fight inside the plant.

"The Manganese Story" and "Missile Watch" are among a number of episodes that join Nelson's heroism with the military importance of the undersea. In "The Manganese Story" (season 1, episode 39), the last episode of the season, Nelson helps two prospectors for "manganese ore, the source of a valuable metallic element vital to national defense" (ca. 1:20). In the episode's climax, one prospector betrays the other and then starts to shoot at Nelson when he surfaces. As is the case with "Flooded Mine," Nelson's immense underwater capabilities save the day. He dodges bullets with "an old frogman trick" (ca. 22:33), leaving the villain "a target to shoot at . . . bubbles of air from my diving lung" (ca. 22:39). He freedives to the boat, then saves the prospectors in a dénouement enhanced by circling sharks. "Missile Watch" (season 3, episode 8) shows how cameras track a missile throughout the course of its launch. When the cameras fail as a result of a US enemy hacking into a communications cable lying deep on the seafloor, the climax of the episode is an extended underwater fight scene between Nelson and enemy scuba divers, prefiguring the hose cutting and wrestling of the James Bond action scenes underwater, discussed in part 3.

The analogy between exploring the undersea and the cosmic frontier of space is another opportunity for technical exposition, from Nelson's comparison of the oxygen used by the test pilot to the tanks of the scuba diver in "Sixty Feet Below." As Nelson explains in "Diving to the Moon," "two strange worlds are opening up to all of us today. One is the world beneath the sea. The other is the alien world of outer space. Strange as it may seem, there are many things that both these worlds have in common" (ca. 0:37). "Diving for the Moon" (season 2, episode 7) starts with a comparison of the two environments, when Nelson uses undersea conditions to test candidates applying for astronaut training. In this episode, however, the promise to show us similarities between either the environments or human exploration there is not fulfilled. Instead, a hurricane thwarts the testing, and the episode takes shape around environmental suspense: "the conservation of our limited air supply was now," in Nelson's words, "the difference between living or dying" (ca. 13:18).

The analogy between exploring the undersea and the cosmic frontier of space similarly offers an opportunity for technical exposition that is then abandoned in "The Aquanettes" (season 4, episode 12), a tongue-in-cheek episode about training female "astronettes" to live in extreme undersea conditions for a trip to Venus, "a world of vapor, mist and water" (ca. 0:33) (fig. 2.33). After starting with the test of spacesuits and voice communication equipment, this somewhat diffuse program turns to the age-old trope of feminine wiles, including flirtation with Nelson and a cat fight among the trainees, until bathing beauties encounter the danger of a shark. Reminiscent of the portrayal of Lotte Hass, gender stereotypes in "The Aquanettes" are vindicated, showing that women do not quite have "the right stuff," despite their impressive credentials given at the episode's beginning. In "Female of the Species" (season 1, episode 6), Nelson is a soft touch for a vulnerable yet determined woman under the spell of her lover in jail, pressuring Nelson for scuba lessons, it turns out, so that she may recover stolen diamonds at the significant depth of 150 feet. In "Cross Current," Nelson searches for a glamorous female marine scientist who has disappeared when she strikes out on her own to investigate submarine currents (season 3, episode 19).

FIG. 2.33. Leon Benson, dir., *Sea Hunt*, season 4, episode 12, "The Aquanettes," ca. 16:32. Underwater training of female astronauts (called "astronettes") in preparation for a voyage to Venus, "a world of vapor, mist and water" (ca. 0:33).

Underwater photography and movie making appear among the many types of submarine activities providing material for adventure in *Sea Hunt*, as the series sometimes took a self-reflexive turn. In "Water Nymphs" (season 3, episode 2), Nelson, photographing fish, is distracted into taking pictures of attractive women who prove to be models in a bathing suit fashion shoot on the surface, which gives the fashion designer the idea to "pivot" "our entire advertising campaign" around "underwater photographs" (ca. 5:51). The episode veers around in a manner similar to "The Aquanettes," as if the series writers do not know how to integrate women into undersea plots, whether they are adventurous or, like the models, more conventionally glamorous. First the models learn to dive, and then the episode settles on style piracy, culminating in the rescue of Inez, a model marooned on a rock at a high tide, by Nelson, requiring a swim to shore underwater, and the arch use of buddy breathing. We learn about movie making in "Monte Cristo" (season 2, episode 5), featuring a director who calls in Nelson in order to use the new capabilities of underwater film to show Alexandre Dumas's hero, Edmond Dantès, escaping

FIG. 2.34. Leon Benson, dir., *Sea Hunt*, season 2, episode 5, "Monte Cristo," ca. 10:03. Mike Nelson on set preparing the stunt dive of "Monte Cristo."

from the burial shroud that is his path out of the seemingly impregnable Chateau d'If (fig. 2.34). At the insistence of the lead actor, Nelson signs off on letting him do the stunt, and instead takes on the role of safety diver. While the sequence is thwarted by the star's jealous stunt double, the stunt would in fact be used in *Monte Cristo* films starting with the 1975 version directed by David Greene.

Sea Hunt also used the sea's perpetually charismatic megafauna to add intrigue. Sometimes, marine creatures feature as attractions; sometimes, knowledge of their habits shapes the adventure plot. "The Mark of the Octopus" (season 1, episode 4), for example, asks Nelson to solve the mystery of a killer whose victims bear suctionlike marks suggesting an attack by an octopus. Yet in this episode, we learn at the conclusion that octopuses are shy creatures, and the telltale markings were inflicted by a villain covering his tracks. In "The Shark Cage" (season 1, episode 13), as well, a fabled marine predator serves as a decoy. In the words of Nelson's voiceover, "a man can be a dangerous creature too and sometimes more treacherous than a shark" (ca. 6:00). The plot of "The Shark Cage," while setting up to focus on sharks, instead turns out to revolve around a man driven to attempt murder out of jealousy incited by his coquett-ish wife. The climax of this episode accordingly returns to environmental sus-pense rather than to solving the problem of danger from sharks. When the villain cuts the cables of a shark cage, sending it plummeting into deep water, Nelson devises a rescue for divers trapped 150 feet below in a cage that had become an "underwater prison" with only half an hour of air.

The narrative patterns of underwater adventure pioneered by *Sea Hunt* proved so robust that they have gone on to form the matrix of many subse-quent underwater adventure sequences in TV and film. Like *Sea Hunt*, these adventures communicate technical knowledge about a remote environment of the planet, and the diverse human activities associated with its application. They organize this information as challenges and solutions, creating environ-mental suspense through limited air supply and the dangers of its toxic atmo-sphere. Additional sources of appeal are the visual exhibition of an environ-ment that few can access, along with charismatic fauna and athletic, attractive divers. *TV Guide* may have been as much mocking as marveling at the aquatic panorama unfurled in *Sea Hunt*'s "epic so watery" (to cite the *Guide*'s sly phras-ing), when it claimed "that Lloyd Bridges' colleagues tell him they have to drain their TV sets after watching his show."[158] Yet while certainly *Sea Hunt* is popular entertainment, the series shares with the genre of the epic a unity of place, action, and characters afforded by an enveloping world, rather than an individual. As in epic, too, the characters and situations across the series affirm the existence and qualities of the realm where they act. Thus, like an epic hero, Mike Nelson embodies the capacities demanded by the cosmos that is his home, which are the reaches and depths of our watery planet.

3

Liquid Fantasies, 1963–2004

With the wet camera unbound in the liquid sky, the submarine environment became a film and TV setting capable of containing a wide range of dramatic imagery and actions. In the latter half of the twentieth century, the wet seduction and sadism of the Bond films, the rapacious killer shark dominating Steven Spielberg's *Jaws* (1975), and the wreckage of a proud civilization in James Cameron's *Titanic* (1997), all created in the aquatic atmosphere, would have an imaginative impact that enshrined these visions as cultural myths. Technical innovations in underwater filming, which filmmakers would continue to develop, made them possible. One emotional condition for the traction of such fantasies was expanding audience familiarity with an environment that, before 1950, most people had never seen. This familiarity was cultivated in film and TV, along with the submarine environment's ever-expanding exhibition across media. Oceanariums and aquariums attracted visitors, and publics continued to expand the pursuit of seaside leisure. A dive media and entertainment complex proliferated, including periodicals, how-to literature, and amateur photography, encouraged by the take-off of scuba and snorkel tourism.[1]

As a result, the underwater environment was becoming by the 1960s a thick and rounded universe in the imagination of general publics, who were familiar enough to infer and to project beyond what was shown. One consequence of such familiarity was that audiences were no longer captivated by traveling shots over beautiful tropical reefs, divers swimming with sharks, and adventures showing the technologies of scuba, even if such views were important to establishing and augmenting the action. At the same time, spectator familiarity with submarine settings released moving-image makers from the imperative to orient spectators and to convince them of the realism of their imagery. As the atmosphere of water continued to yield unique visuals, as well as to shape fresh story lines, the underwater environment opened new possibilities to narrative cinema for fantasy creation.

In part 3, I describe memorable liquid fantasies created on the underwater film set, from the buddyship of boy and dolphin in the film *Flipper* (1963), directed by James B. Clark, to the send-up of *The Undersea World of Jacques Cousteau* as a cultural phenomenon in Wes Anderson's *The Life Aquatic with Steve Zissou* (2004). From *Flipper* to *The Life Aquatic with Steve Zissou*, the films and filmmakers discussed in part 2 leave their impact. As Anderson, among others, illustrates, Cousteau is a beloved reference point. The gothic representation of the shipwreck in *The Silent World*, for example, shapes the submerged ruins of modernity in *Water World* (1995), directed by Kevin Reynolds, as well as in the opening of James Cameron's *Titanic* (1997). The underwater heroism of the dolphin Flipper, appearing first in a film and then in the TV series overseen by *Sea Hunt* producer Ivan Tors, adapts Mike Nelson's underwater feats in *Sea Hunt*. The James Bond films amplify the breezy glamour and exploits of *Sea Hunt* and *Diving to Adventure*. Fleischer's undersea wreck set in *20,000 Leagues under the Sea* turns into an elaborate haunted house in *The Deep* (1977), directed by Peter Yates, which features some of the most extended screen time underwater in the history of narrative film.

From *Flipper* to *Titanic*, I focus on works, mostly commercial successes when they debuted, whose fantasies go on to thrive in the cultural imagination. For example, *Flipper*'s fantasy of friendship, first launched in a film, and then reprised in a three-season TV show by the same name (1964–67) went on to shape numerous interspecies romances with charismatic marine mammals. These romances include in a tragic mode Luc Besson's *The Big Blue* (1988), subsequently discussed, or more affirmingly, the *Free Willy* series (1993–2010), where children befriend an orca, an emblem of pristine nature harmed by humans. The television series *Baywatch* (1989–2001), which the *New York Times* declared in 1995 to have a "wider audience on the planet Earth than any other entertainment show in history" offers good examples of both the afterlife of Bond sensationalism and of *Jaws* horror.[2]

In part 3, as in previous sections, I leave to the side films about submersibles, where so much drama is created in the atmosphere of air. Relatedly, I do not analyze fantasies of the underwater habitat, which emerges as an aspiration for human dwelling in the 1960s.[3] I only take up such settings when they offer the portal to new framings of the depths. This occurs in *The Abyss* (1989), where Cameron immerses the operators of the submersible mobile drilling rig *Deepcore* to reveal aliens inhabiting the remote recesses of the planet.

Some liquid fantasies play out over extensive underwater sequences; a liquid fantasy can also be a brief but memorable flashpoint in a film. I organize the films in roughly chronological fashion to reveal how fantasies transform with film's own intertextuality. For example, the use of the swimming pool to express the warped values of society on land is launched by a shot from the bottom of a pool of a dead body at the opening to Billy Wilder's *Sunset Boulevard* (1950), about Hollywood decadence. In *The Graduate* (1967), Mike Nichols uses the point of view shot from the bottom of the pool to express materialist suburban anomie. Steven Spielberg would add a futuristic twist in the blurry images of drowning seen by the psychic Agatha in *Minority Report* (2002), enslaved in toxic-looking cloudy liquid by a dystopian start-up.

3.1. Symbiosis and Kinship: *Flipper*

An emotional bond that crosses atmospheres is the subject of the film *Flipper* (1963), based on a book by Ricou Browning, directed by James B. Clark, and produced by *Sea Hunt* producer Ivan Tors. Browning, who had starred as Gillman in the *Creature* trilogy, by his own account had used "his last $100 to write a book telling the boy and dolphin story."[4] Browning's gamble paid off, and Tors decided to make the movie, calling on *Sea Hunt* cameraman Lamar Boren. The plot of the movie focuses on Sandy Ricks, played by Luke Halpin, the son of a poor, hardworking fisherman, Porter, who prefers the rigors of coastal life to dull working-class routine on land. Sandy has been taught by Porter, played by Chuck Connors, and his mother Martha, played by Kathleen Maguire, to practice a can-do, Christian compassion that leads him to befriend a wounded dolphin. As Sandy's friendship blossoms with Flipper, played by a dolphin named Mitzi, the boy leads his tough but caring father to give up his belief that dolphins are a fisherman's competition. Rather, Porter learns thanks to his son, the dolphin is humanity's friend and fellow creature; in the father's words, Sandy leads him to "an understanding of their kind" (ca. 3:04).

Gregg Mitman reveals the media and science complex that constructed dolphins as "playful, communicative, highly intelligent creature[s] of the sea" in the popular imagination during the 1950s.[5] The first dolphin star in the famous oceanarium Marine Studios, located in Florida, was "Flippy, the educated dolphin," a two-year-old male caught in 1950 in a "nearby estuary" and then extensively trained. Mitman emphasizes how oceanarium stars with names like Flippy, or "Grumpy, the loggerhead turtle," reinforced the domestication of undersea creatures and that naming animals had "a long tradition within natural history film."[6] In the 1950s, in Mitman's assessment, the dolphin found "its popular image shaped . . . by the constructed ideals of family life."[7] Mitman mentions documentary film that played a role, notably the 1959 Disney True-Life Adventures' *Mysteries of the Deep*, filmed at Marine Studios, showing the "miracle of living reproduction" and the care of the baby by the maternal female dolphin.[8]

Even as the film *Flipper* was inspired by human-dolphin interactions in oceanariums, we can see its visual innovation taking spectators beyond the limits of the oceanarium exhibition format. One limit had to do with the difficulty witnessing action. Marine Studios inaugurated two types of viewing experiences that would become essential to the oceanarium's allure. The first was the topside view of tanks, and the other was the views of marine life through windows on the depths. Indeed, when Marine Studios opened, its designer, William Douglas Burden, realized the affinity of the underwater view with the absorptive experience of cinema. Burden decided to design the oceanarium so that, in his words, the "usual distractions that are so ever-present in the exhibition hall of a museum or aquarium" were absent.[9] One needed to create "the conditions of the motion picture theater," and Burden even imagined that the viewers would be screened on either side by a curtain so as not to be disturbed while enjoying this peep into the sea.[10] The topside view from the tanks, however, became just as if not more important in the oceanarium format, which included the exhibition of trained marine mammals like Flippy performing entertaining tricks.

Both views were exciting; however, spectators could not be in two places within seconds. A favorite dolphin feat in oceanariums is for the dolphin to leap entirely out of the water. The leap shows the dolphin's athleticism and it also satisfies briefly the desire of the spectator watching from above to see the whole dolphin. With the technique of montage, topside views of Flipper cavorting, which show primarily the fin, tail and head, can transition within less than a second to an underwater view that makes the rest of the body visible. Such montage of topside and underwater views of Flipper swimming and performing recurs throughout the film, characterizing the playful, friendly bonding interactions between Sandy and Flipper.

The one-hour mark is an important emotional moment in a film. One hour into *Flipper*, Sandy shows off Flipper's intelligence and tricks by creating a home oceanarium in his father's fish pen. The children gather round, delighted, as Sandy has Flipper fetch a hat, his swim gear, and other objects, to the accompaniment of the song "Flipper," by Henry Vars and By Dunham, a lilting,

FIG. 3.1. James B. Clark, *Flipper*, ca. 1:28:34. Sandy riding the dolphin Flipper—part bucking bronco and part age-old companion (see fig. 3.2).

childish melody set to the ¾ beat of the waltz. The amateur oceanarium highlights the film's liquid fantasy and its message that humans and dolphins are kin and transcend the separation between the elements of air and water.

Terrestrial humans' desire to overcome the separation between creatures of the air and sea is the subject of a long-standing mythology reaching back to the classical era and stretching into our own (from the myth of Glaucus, for example, and the Selkies, to charming children's books such as Leo Lionni's *Fish Is Fish* [1970]). Sandy's working-class father surprisingly cites a story from Herodotus, which Porter learned from the Greek fisherman Nick, who drowned in the hurricane that starts the film. The story recounts the fate of the poet Arion, condemned to die on a shipboard journey by a crew that covets the wealth he has won singing. When the crew throws Arion overboard, he is rescued by one of the dolphins who have come to hear his sweet song. First Sandy and then Porter return to the story in the film's conclusion, when Porter agrees that "it was written for us today to share with the dolphins, not to kill them" (ca. 1:28:23).

The father's pronouncement transitions to an ending sequence showing the dolphin topside and underwater with Sandy riding on his back (a stunt that the teenage actor learned from Browning) (fig. 3.1). With Sandy's hand held up as if he is on a bucking bronco, albeit awkwardly, his pose suggests a child playing at the Wild West, in contrast to the elegance of the subject in its

FIG. 3.2. Piazzale delle Corporazioni, detail from *Trasporti Navali* (sea transport), mosaic, Ostia Antica, Italy. Licensed under CC BY-SA 3.0. Credit: sailko. The classical poet Arion similarly rode on the back of a dolphin, according to a story Sandy's father hears from a Greek fisherman and recounts to Sandy.

classical representation, as can be seen in an Italian mosaic from Ostia (fig. 3.2). The topside and underwater views of this scene contrast as well with topside/underwater montage of the dolphin's spectacular swimming that dates to Cousteau and Malle's *The Silent World*. In *The Silent World*, the dolphins are majestic, and the seamen marvel at a distance. *Flipper*, in contrast, folds topside and underwater views into the circle of the nuclear family.

In *Flipper*, both underwater and topside views are generally provided by a spectator point of view exterior to the scene. But there is one remarkable sequence preceding the home oceanarium scene when spectators are invited to inhabit the perspective of the dolphin. In this perspective, we see Sandy through the water's surface, in a dinghy, looking down with delight. Shot with a fish-eye lens, this submerged view is shimmery, distorted, and yet distinct. Sandy's stretched image undulates with the movement of water (fig. 3.3). The sequence repeats Flipper's liquid perspective of Sandy several times, matches this perspective with an underwater shot of Flipper's eye, and further incorporates views of Sandy in the atmosphere of air, looking down at Flipper from the dinghy with a big smile (ca. 56:28). In contrast to *The Silent World*'s

FIG. 3.3. James B. Clark, *Flipper*, ca. 56:32. The liquid perspective of Flipper.

enigmatic depiction of the relation between Jojo and Delmas, Sandy and Flipper are suggested to exchange gazes, and indeed, the next underwater sequence shows the two swimming together, with Sandy petting Flipper.

The *New York Times* acknowledged the skill that went into creating what it called this "simple boy-meets-dolphin affair." In praising "beautiful views of marine nature, big fish, little fish, turtles and skin divers," the reviewer wrote that "the main credit," "seems to go to the cameramen" to the boy and dolphin, to those who trained her—and of course, "to producer Ivan Tors, a man who has done so much subaqueous filming that he undoubtedly works out his budget with pens that write under water."[11] Tors would go on to produce a three-season, eighty-eight-episode TV series, *Flipper*, inspired by the movie, with some of the original film's production crew. Browning wrote thirty-three episodes, and Lamar Boren served as underwater director of photography for twenty-eight episodes in 1966–67. In between the film *Flipper* and the TV episodes, Browning and Boren collaborated on the underwater sequences for the first James Bond film to plunge beneath the surface, *Thunderball* (1965).

3.2. The Underwater Peep Show: *Thunderball*

When producers Albert Broccoli and Harry Saltzman began the fourth James Bond film, *Thunderball*, directed by Terence Young, they envisioned offering audiences an intense sense of documentary reality. Ricou Browning recalled

that at the project's beginning, Broccoli and Saltzman sought to "sign up underwater legend Jacques Cousteau," because they "wanted . . . realistic" underwater filming.[12] Cousteau, moreover, was friends with James Bond's creator,
Ian Fleming, who had first learned to scuba dive from Cousteau in 1953. Fleming had become interested in frogmen and underwater espionage while working at the British Department of Naval Intelligence during World War II.
Among the "ruses de guerre" suggested by the Department were "schemes to
lure U-boats and German warships towards false shipwrecks surrounded by
mines, and spreading the (serious) rumor that turtles had been sighted around
the British Isles."[13] Fleming wrote newspaper articles about underwater exploration, covering subjects such as treasure hunting, Cousteau, and diving himself, describing the "odd sensation of being mistaken for a shark by a Remora
fish, which attempted to attach itself to his body in the hope of scavenging a
meal."[14] Fleming first used scuba in the novel *Live and Let Die* (1954), drawing
not only on Fleming's experience but also the "real-life wartime adventures of
the Gamma Group, crack Italian frogmen who sank over 54,000 tons of shipping in Gibraltar."[15]

However, instead of Cousteau, the producers hired Browning, who offered
a canny assessment of his value. "Cubby [Albert] Broccoli asked why you,
rather than him [Cousteau]. I said: 'For one reason. We are phonies. Everything we shoot is phony, everything he shoots is live, so we can make it better.'"[16] Along with Browning, Broccoli and Saltzman hired Boren, who would
do the underwater camerawork for subsequent Bond films produced by Eon,
including *You Only Live Twice* (1967), *The Spy Who Loved Me* (1977), and
Moonraker (1979—uncredited). *Thunderball* had a $9 million budget, which,
to give a sense of proportion for 1965, was "three times that of *Goldfinger*."[17]

Cousteau's prominence in underwater capture notwithstanding, the film's
aspects that pleased viewers, if we take the opinion of Bosley Crowther, are
not exquisite documentary but rather "diving saucers, aqualungs, frogman
outfits, and a fantastic hydrofoil yacht" as "devices of daring and fun."[18] His
appreciation of *Thunderball* centers on how the film plays with cool aspects of
diving and water sports culture. "The amount of underwater equipment the
scriptwriters and Mr. Young have provided their athletic actors, including an
assortment of beautiful girls in the barest of bare bikinis, is a measure of the
splendor of the film."[19]

Thunderball developed its aquatic spectacle with a royal budget and wide-
screen color format, putting *Sea Hunt*–style combat on steroids. The designers
had spray-painted Bridges's wet suit silver so he could be picked out; in

Thunderball, the team of good guy divers are clad in orange, while the bad guys' wet suits are black. Hose-cutting and brawling recur throughout underwater fights, and Bond even uses the "old frogman trick" (Mike Nelson) in the underwater finale, when Bond decoys Largo's men with bubbles from his compressed air tank that he then ditches. Nonetheless, *Sea Hunt's* spare action had now yielded to scenes choreographed with "sixty divers using more than $85,000 worth of diving equipment."[20]

The equipment in the Bond films tickled audiences, themselves joining an epidemic of what the *New York Times* called "gadgetitis," by which was meant all manner of "gaily colored gadgets" ornamenting summertime beaches.[21] The exuberant Bond gadgetry even attracted the professional attention of the Royal Corps of Engineers, who queried designer Peter Lamont about the emergency tiny rebreather invented by Q. How long did it last, they wondered, to which Lamont apparently replied, "As long as you can hold your breath."[22] The most famous underwater Bond gadget would appear twelve years later in *The Spy Who Loved Me* (1977), directed by Lewis Gilbert; nicknamed Wet Nellie, the car was bought by Elon Musk in 2013. As the *Telegraph* explained, in reality "Roger Moore never drove Wet Nellie either, of course, and it's unclear whether he even sat in it. The interior shots were of a different car, and the pilot of the vessel was Don Griffin, a retired US Navy SEAL, who operated it wearing full scuba gear with an auxiliary oxygen supply."[23]

The shark scenes in *Thunderball* also took *Sea Hunt* wildlife to the next level. The film had fifteen sharks brought in for the scene where Connery is thrown into Largo's pool. To keep Sean Connery, who was playing Bond, safe, designers relied on a Plexiglas barrier. However, they did not install enough Plexiglas, and sharks passed through the gap. Thus, there is a shot of a terrified Bond in the frame with a shark passing by, where Connery apparently was not acting.[24] For the sharks in the underwater finale, Browning recalled, "we found that the Tiger shark was better to work with in that if you let him go, he would kind of go in a straight line. But if you had a Lemon shark or a Bull shark and you let him go," sometimes they would turn and "they're facing you."[25] Even though this behavior was deemed less menacing, nonetheless, the animals were subjected to such barbaric practices as placing "wires through [their] fins to control their movements, keeping the rest of the crew in relative safety."[26]

Along with the fun praised by Crowther, *Thunderball* used the gadgets and the aquatic haze to suggest darker fantasies. This suggestion starts with the film's mesmerizing credit sequence by Maurice Binder, who created the stylish, abstract Bond credits throughout the 1960s and '70s.[27] For *Thunderball*, information on the film's production is featured against a hunt/dance of athletic scuba divers clad in wet suits and wielding spear guns chasing shapely, naked women swimming gracefully. The actors were shot underwater and, hence, move with a distinctive aquatic glide. They swim, however, through an abstract, vivid sea of bright color, alternating the seven colors of the rainbow. Indeed, in a wink, the only murky green in the vivid credit sequence is matched to information about the film's underwater photography. Even with bright colors, the credits preserve the aquatic haze, which enables naked women, bare breasts moving, to swim before general audiences, in a fashion that would be censored if shot through air. Binder's equivalent to nudity in *The Starfish* or Jean Vigo's shot of Taris's buttocks is the woman who swims from the bottom of the frame to the top with her pubic area pushed into the spectator's face, just before introducing Ted Moore as the director of photography.

The prominence of spear guns in this underwater dance is in keeping with the Bond novels famous for their "[s]ex, snobbery, and sadism," to echo the headline from a 1958 review by Paul Johnson in the *New Statesman*.[28] We have seen films from the 1950s exalt the power of dive technologies and explore the submarine environment's potential for adventure. The underwater sequences in Bond films, in contrast, starting with the credits of *Thunderball*, make a link between diving gear and the materials of fetishistic sex play. Wet suit rubber accentuates the body, and masks depersonalize and empower their wearers, who wield phallic spear guns (the fetish as defined by Freud is a phallic substitute).

The plot of *Thunderball* continues to make explicit the fetishistic possibilities of scuba gear in its transformation of the dive adventuress. In the 1950s, women like Lotte Hass and Zale Parry were wholesome beauties, and light-hearted flirtations were the extent of romantic underwater diversion. In *Thunderball*, in contrast, the dive adventuress turns into a masochistic partner, tellingly named Domino, subjected both to her "guardian," Emilio Largo, and, more enjoyably, to Bond. Domino was played by Claudine Auger, who was doubled underwater by Evelyne Boren, wife of Boren. When Domino is first introduced, her garb and pose pay homage to Lotte Hass. In *Under the Red Sea*, we first see Hass exploring the reef, in a belted black bathing suit, with streaming long hair. Domino is introduced in the same pose, before she gets her foot wedged in a rock and has her first encounter with Bond (figs. 3.4, 3.5). After this meet cute, the two will flirt on land under the watchful eye of Largo until they briefly escape for an underwater sex scene, featuring Bond and Domino's

FIG. 3.4. Hans Hass, *Under the Red Sea*, ca. 11:16. Lotte Hass exploring the reefs.

FIG. 3.5. Terence Young, *Thunderball*, ca. 44:34. Domino's first appearance—in the same pose as Lotte Hass and wearing a similar swimsuit.

FIG. 3.6. Terence Young, *Thunderball*, ca. 1:37:24. Rubber embrace.

rubber embraces, with a tongue-in-cheek climax off-screen in the depths (fig. 3.6).

When the members of the British Board of Film Classification (BBFC) were first shown the script of *Thunderball*, they had doubts that the film would qualify for an "A" certificate for general audiences. In the words of "John Trevelyan, of the BBFC, 'I get the impression that this screenplay has been deliberately hotted up with a view to its including more sex, sadism and violence than the previous Bond pictures, and . . . it seems less light-hearted in tone.'"[29] The BBFC noted over thirty details in the script that needed to be "substantially modified." Among underwater scenes to remediate, the board flagged Bond's discovery of the dead crew in the NATO plane, hijacked by Largo for its warheads. In addition to warning against the display of bodies, the board also instructed the producers not to have sharks devour corpses; "we would not want to see a lot of blood stained water."[30] When the BBFC examined the final version of the film to see if it was acceptable for general audiences, the only cut required was of Bond "stroking the back of a partially nude girl with a mink glove."[31] The gory killings in the final epic underwater battle scene were retained, which suggests that the BBFC's instructions concerned form more than substance. Or perhaps the hazy atmosphere dilutes the emotional force of the gore, just as it discreetly veils provocative nudity. In stark contrast to the BBFC instructions apropos of one of the final fights—perhaps the fight onboard—that "it should be short and not too brutal," the underwater fight scene goes on for over eight minutes—truly an extensive amount screen time underwater in the history of underwater film—from 1:54:16, when the first

FIG. 3.7. Terence Young, *Thunderball*, ca. 2:01:16. Underwater gore.

parachutists hit the water, to a few moments after 2:03:00, when Bond chases Largo into the *Disco Volante*. Rather than suspense about Bond's survival, the scene solicits our enjoyment of wet suit–clad men locked in combat, amidst clouds of water and blood (fig. 3.7).

In her famous article on the gendered divide of the gaze in the pleasure we take in watching cinema, Laura Mulvey suggested that woman is the object, fetishized in the glamorous stars of Hollywood films from the 1940s, while man is the looker, since "[m]an is reluctant to gaze at his exhibitionist like."[32] When women are eroticized as objects of the gaze, they often wear parapher- nalia recalling the phallus to accentuate such fetishism (Freud identifies the display of the phallus as the visual focus of its pleasures). When Honey Ryder, played by Ursula Andress, makes her entrance in *Dr. No* (1962), the first Bond film, she emerges from the water in a bikini, with a knife hung from a belt at her hip. "Looking for shells?" Honey asks Bond (ca. 1:03:11). "No, I'm just looking," he says, articulating the pleasure identified by Mulvey.

In *Thunderball*'s underwater fight scenes, the men are clad in rubber and adorned with dive prostheses. They qualify as what Amanda Feuerbach calls the "hypermasculine cyborg," whose invention she credits to recent science fiction, such as *Terminator 2: Judgment Day* (1991). The core features that make up the hypermasculine cyborg for Fernbach are "phallic props," which are "the equivalent of the Hollywood female star's accouterments—highly stylized and flawless makeup, feathers, stockings, and so on."[33] For the cyborgs in *Termina- tor 2*, accoutrements include black leather, a motorcycle, and technobody parts. Scuba gear and submarine weaponry played this role in *Thunderball*

FIG. 3.8. Terence Young, *Thunderball*, ca. 1:06:27. The male star as object of the gaze.

almost thirty years earlier. Both *Terminator 2* and *The Terminator* were directed by diver-director James Cameron, and in between these two films, Cameron made *The Abyss* (1989).

Thunderball further lets us gaze at men in fetishistic water gear in the atmosphere of air, albeit somewhat veiled by darkness. Bond, following an escape from nighttime underwater spying on Largo, emerges glistening wet from the sea. In this shot from the sequence (fig. 3.8), Bond is in the process of taking his infrared camera from around his neck, signaling that he is no longer gazer but object. He looks down, rather than confronting the spectator, a direction that when used for female figures in art, emphasizes their object status. He is wearing a beavertail wetsuit, fastened at the bottom, which draws attention to his crotch and shows off his torso. A skin diver's knife strapped to his shapely calf echoes the appearance of Honey Ryder. The gleams of light on his body, shiny from the water, catch the eye. The gleam has been associated with fetishism by Freud, concentrating erotic energy. Bond emerging from the water at night is a male instance of a "visual presence [that] tends to work against the development of the story line, to freeze the flow of action in erotic contemplation" that Mulvey identified as a quality of the female star in Hollywood conventions.[34]

The underwater atmosphere as an occasion for voyeurism of both female and male stars would become a staple of James Bond films from the fourth film, *Thunderball* (1965), through to *The Spy Who Loved Me* (1977), *Moonraker* (1979), *For Your Eyes Only* (1981), and *Never Say Never Again* (1983). The Bond style of underwater peep show would also appear in other spy thrillers and franchises, such as, at the turn of the millenium, *Mission: Impossible* and the Bourne films.

3.3. Horror Points of View of the
Apex Predator and Its Prey: *Jaws*

In part 2, I discussed the opposing visions of sharks that were put forth by Hass, and Cousteau and Malle, in the 1940s and 1950s. For Hass, sharks were majestic inhabitants of the deep, to be respected and studied. Cousteau would come to this more symbiotic environmental view, but in *The Silent World*, he and Malle emphasized the age-old terror of sharks, and shark carnage. As I mentioned, the film's shot/countershot formula would be taken up in a lineage of narrative films representing sharks, starting with *The Old Man and the Sea* (1958). The technique would come back to its horror home when Steven Spielberg merged *Moby-Dick* with horror conventions and created a blockbuster, "the first movie released in more than 400 theaters in the United States, and the first movie to gross over $100 million at the box office."[35]

Critics have discussed the horror conventions of *Jaws* and its resemblance to a submarine film and to *Moby-Dick*, among a number of other topics.[36] Of particular note for the history of liquid fantasy are the ways Spielberg adapts and amplifies Cousteau and Malle's poetics of shark carnage into a two-hour narrative. For Cousteau and Malle, who invented this poetics, imagery of sharks feeding revealed vicious predation. Boren had alternated topside and underwater views to similar effect in the brief shark sequences taking the viewer underwater in *The Old Man and the Sea*. Another notable antecedent was the 1971 *Blue Water, White Death*, directed by Peter Gimbel and James Lipscomb, about documentary filmmakers who set out, led by Gimbel, to find and film a great white shark. Gimbel was friends with Peter Benchley, the author of the novel *Jaws* (1974), and the final sequence of the documentary, when the great white attacks the cage of photographer Peter Lake and deforms the bars, was one inspiration for the novel. Indeed, Australian shark experts Ron and Valerie Taylor, who star in *Blue Water, White Death*, provided footage for the sequence when the shark rams the cage in the final encounter of *Jaws*. Further, there are echoes of Gimbel in Hooper, the independently wealthy marine biologist.

In *Blue Water, White Death*, the topside/underwater montage is gripping, playing off the contrast between the views of the shark as they appear from the boat and those that are revealed in the depths. There is nonetheless a playful atmosphere to the film, which the *New York Times* film critic Vincent Canby described as "the quite jolly, sometimes awesome, new documentary," fitting within the "conventional documentary of romantic experience."[37] The adventurous and knowledgeable team of underwater filmmakers—Stan Waterman, Ron and Valerie Taylor, and photographer Peter Lake—keep it cool, even as the excitable Peter Gimbel remarks dramatically on the extraordinary quality of their adventure.[38] The soundtrack of the film is equally low key, mostly natural sounds corresponding to the setting and action, with some chill music, such as Tom Chapin singing folk songs on deck. Canby grasps the "heart of the film," which is "its action recorded with immense technical skill . . . so pure that it's as poetic as anything I've seen on the screen in a long, long, time."[39] He singles out "some of the most beautiful and (there is no other way to describe it accurately) breathtaking underwater footage" of sharks.[40] This imagery is still a point of reference today—notably in assessing subsequently despoiled habitats, where shark numbers have dwindled. The awe-inspiring views of sharks at home in their element makes the surface thrashing of the poetics of shark carnage devised by Cousteau and Malle seem sensationalist. Even in the final episode, when the great white shark butts, deforms, and almost enters Peter Lake's cage, the mood does not turn against the massive predator and portray it as a villain. Rather, the participants laugh and make light of their encounters.

Where *Blue Water, White Death* took viewers along on a film safari, Spielberg invites them to a two-hour spectacle of suspenseful horror. Much of *Jaws*'s suspense derives from the contrast between topside opacity and underwater carnage invented by Cousteau and Malle, beginning with the film's opening montage. We first see a beautiful girl, Chrissie, played by Susan Backlinie, skinny dipping and swimming out to a buoy at dawn. The water is still, and the scene idyllic, as she sinks underwater, leaving a taut leg exposed for spectators to admire, like a dancer, toes pointed. And then, we get a view looking up from the depths of her smoothly swimming silhouette.

Whose view is this? As Andrew Gordon rightly observes, the stalker conventions in *Jaws* have an underwater precedent from the dawn of Hollywood's use of scuba for shooting in *Creature from the Black Lagoon* (1954). That film features a scientist's assistant, Kay, played by Julie Adams, stalked by the Creature below.[41] Kay is an elegant swimmer, and the first underwater view of her taking a dip in the lagoon enables us to admire her stroke. As the film connects her surface beauty with her underwater glide, the camera sinks deeper to the level of some seaweed, and spectators receive a shock when it reveals the back of the head of Gill-man, and then switches to his more vertically angled view of Kay. The menace revealed in the deep inaugurates a memorable suspense sequence, as a distanced, underwater point of view shows us carefree Kay

FIG. 3.9. Jack Arnold, *Creature from the Black Lagoon*, ca. 30:12. Horror pas de deux.

enjoying the water, with the creature following beneath her (fig. 3.9). We fear he will kill her, but in an added twist, Gill-man falls in love and abducts her.

When Spielberg echoes this scene in *Jaws*, he does not reveal the source of the underwater gaze. Rather, he takes up a shot from *Creature*, of Gill-man viewing Kay's legs, and lingers on Chrissie's legs seen from below, accompanied by the film's unforgettable theme, composed by John Williams. In Spielberg's version, the shark-eye point of view of human legs dangling in the water becomes an iconic shot (fig. 3.10).

Spielberg repeats the montage of shark's-eye view and a topside camera's image of swimmers throughout the first half of the film, including in several attacks. In these subsequent scenes from the beach, Spielberg drops Chrissie's grace and uses the dangling legs of holiday beachgoers of all ages to express menace lurking in the waters. He maintains suspense by varying the victim in the first masterly beach scene of destruction (Pippin, the dog; a child on a raft), and by hiding the moment when the shark strikes. One of his brilliant innovations throughout this first half is to sever the connection that Cousteau and Malle, like subsequent filmmakers inspired by their conventions, create between the underwater and surface perspectives. Throughout the first part of the film, we never see the shark's form underwater, or its appearance to

FIG. 3.10. Steven Spielberg, *Jaws*, ca. 4:40. Shark's-eye horror point of view—but no view of the shark from the spectators' perspective.

people at the surface. Instead, Spielberg keeps topside and underwater views apart for as long as the plot can delay the connection.

Gordon explains that Spielberg, to enhance audiences' engagement, added a third type of shot to the contrast between underwater visibility and topside obscurity: "a look for *Jaws* that simulated a person's point of view while swimming."[42] Cinematographer Bill Butler designed a special platform with a section cut out to accommodate the camera in its water box, enabling shots around the waterline.[43] With this camera, Spielberg was able to create a shot that expresses the viewpoint of the swimmer, imbued with a sense of menace from the unseen depths. In psychological terms, he "plays upon the viewer's wish to know and yet not to know the monstrous, to see and yet not to see."[44]

Jaws uses such surface views in its sequences of shark carnage from Chrissie's death—and withholding the connection of such views to underwater action enhances their impact. Critics compare this opening scene to the shower scene in Alfred Hitchcock's *Psycho*. Gordon, for example, notes the common elements: "the sexual frisson of the beautiful naked blonde, mixed with suspense and terror at her vulnerability, and the killer, announced by a point-of-view shot."[45] The agonized struggle of the victim and the spectator's thwarted desire to intervene further intensifies the sense of horror. Chrissie does not sink into the depths immediately; rather, she flails, caught by the hidden monster as her suffering is amplified by the obscurity of its source. In contrast to the "rapid, shock cutting" used in *Psycho* and *The Silent World*, the camera records this flailing coldly from the outside, letting the actress do all the work in "relatively long takes" where "[t]he camera moves very little; rather the victim is dragged back and forth across the screen."[46]

This surface view recurs both in instances of death—as in the case of the boy on the raft in the first sequence of beach carnage—and in the instances of swimmer panic at the dramatic marker an hour into the film. In this scene, the camera just below the waterline also shows us the terrified faces of bathers above the surface, wondering whether the shark is below them. Along with its link to the perspective of ordinary viewers, Spielberg's focus on the surface of the sea draws emotional potential from a zone that was underutilized in the first era of scuba filmmaking, which focused on the depths. Ann Elias notes the visual and epistemological fascination of the sea's surface, what Rachel Carson called its "mantle" in *The Sea around Us*, and Elias writes arrestingly of its threatening fascination, when she discusses director Frank Hurley's attempt to look beneath the surface at the Great Barrier Reef. "In Hurley's photograph[s]," Elias writes, "it is not the ocean's depths that are sublime but the skin of the sea and the vision of corals lying half-submerged and half-revealed by water, signaling menace, danger, peril and wreck."[47] Elias's interpretation translates well to Spielberg's cinematic appreciation of "the harrowing transition zone," a phrase Elias takes from Barry Lopez.[48]

Spielberg prolongs the disjunction of the topside and underwater views for almost half the film. A little over a full hour elapses before the two zones are joined together, as the shark breaks the surface and looms there for a brief moment when it strikes (fig. 3.11). Lore has it that one practical reason

FIG. 3.11. Steven Spielberg, *Jaws*, ca. 1:03:26. Finally, we see the shark, when it breaks the surface at just about the one-hour mark of the film.

for this delay was Bruce, the production's mechanical shark, which rarely worked, in particular in the open water where the film was shot. In any case, the shark's appearance enables Spielberg to add some adventure narrative to what is predominantly a horror film. Even as a lengthy shark hunt takes up the second half of the film, the film continues to use topside and underwater views, sometimes stitched together, sometimes separated, for horror suspense. At the film's climax, the surface view shows the shark sweeping away Quint, played by Robert Shaw, awash in gore and thrills.

Spielberg's film amplifies the image of sharks in previous documentaries, even those with violent imagery, into a malevolent killer so predatory that no human is safe. As Noël Carroll observes of *Jaws* and its sequels, "the sharks . . . seem too smart and innovative to be sharks . . . capable of carrying out long term projects of revenge way beyond the mental capacities of its species. Indeed, the shark in these films manages to kill about as many humans in a single summer as all the actual sharks in the world do a year [in fact more]. These are not the creatures of marine biology but fantasy."[49]

That fantasy captured audiences' imagination. Richard Ellis notes that "ever since the 1974 publication of *Jaws* and the hugely successful movie directed by Steven Spielberg, authors have tried to improve on Peter Benchley's . . . formula," notably by making the monster bigger and, if possible, more evil.[50] Both the malevolent shark and Spielberg's prolongation of the separation of underwater and surface views have been taken up by shark spectacles in both literature and film, most recently in summer thrillers like director Jaume Collet-Serra's *The Shallows* (2016) and John Turtletaub's *The Meg* (2018). This fantasy appears as well in countless TV shows, including the Discovery Channel's annual summer TV "shark week," which promised in 2019, "A Summer Event of Bigger Sharks and Bigger Bites."[51]

The poetics of shark carnage stretched out to feature-length predation is Spielberg's principal contribution to liquid fantasy with *Jaws*. Yet, as a master of horror, Spielberg realized the potential of shipwreck as another opportunity for this aesthetic. He started to develop this potential in just one sequence set at night, using darkness and fog to create a mood of ominous foreboding. The shark biologist Matt Hooper, played by Richard Dreyfuss, goes out with Chief Martin Brody, played by Roy Scheider, to spot the shark, which is a night feeder, Hooper tells Brody. Hooper free dives to investigate a capsized boat that, he realizes, belongs to a missing fisherman. Surveying the overturned craft underwater, Hooper discovers a shark tooth in its side. As Brody continues his examination in the murky atmosphere, all the more mysterious because it is night, he suddenly sights the blanched head of the missing fisherman transfixed in a death stare, with one eye ripped out. A scream divorced from any character in the scene accompanies the jump scare. This fantasy worthy of a horror house contrasts with the respect for the tragedy of wreck sites in Cousteau's *Sunken Ships* and Cousteau and Malle's *The Silent World*. DeMille, too, had given the wreck set a mournful tone in *Reap the Wild Wind*. *20,000 Leagues*, in contrast, started to approach the interest of exploring its recesses in the search for sunken treasure, although the wreck set only featured briefly in the film. Both the interest of sunken treasure and the use of the wreck site as a horror house were important in the film adapted from the next novel Peter Benchley wrote after *Jaws*: *The Deep* (1976). Peter Yates promptly adapted the novel into a film; as *Variety* declared, *The Deep* (1977) was "a clear attempt to cash in on the success of . . . *Jaws*."[52]

3.4. Treasure Hunting amidst Underwater Moods: *The Deep*

The Deep depicts the adventures of amateur treasure hunters looking to get rich through wreck diving in the Caribbean. An American couple on vacation, Gail Berke and David Sanders, played by Jacqueline Bisset and Nick Nolte, stumble on a sunken World War II ship with a cargo of morphine, which also turns out to hide a Spanish wreck from two centuries before. Both finds capture the interest of locals. The lure of morphine intrigues the local drug dealer, Henri Cloche, played by Louis Gossett Jr. The lure of Spanish treasure draws into the plot Romer Treece, played by Robert Shaw, the actor who played Quint, the vengeful shark hunter, in *Jaws*. In making this film, Peter Guber hoped to produce a "film richer in underwater visuals than has ever been done before . . . from close-ups of golden treasure gleaming in underwater sand drifts to panoramic vistas of a huge shipwreck sprawled across the ocean floor."[53]

The production team sought to create "all the spectacular 35mm Panavision visuals which feature film viewers have come to expect from feature films shot on land—underwater," as well as to apply them to specifically aquatic vistas and fantasies: "the majesty and terror of the deep, portrayed in exquisite detail and jewel-like color."[54] With a budget of $8.6 million, they had the resources to try. Producers hired as directors of underwater photography Waterman, who among many credits, had produced and filmed *Blue Water, White Death*

(1971), and the up-and-coming Al Giddings, who had documentary experience and had just shot the narrative film *Sharks' Treasure* (1975), directed by Cornel Wilde. Giddings designed three 35mm cameras for underwater filming and declared his aim was "to shoot . . . as a creative topside Director of Photography would": "with different focal lenses, three cameras shooting simultaneously . . . [to] give the editor something really dramatic to cut."[55] The underwater cinematography team included as well photographer and cameraman Charles Nicklin, who has entered into the International Scuba Diving Hall of Fame, along with star still photographer David Doubilet.

In making *The Deep*, filmmakers used ocean settings for orientation and montaged these views with action footage made in more controlled studio sets. Giddings disclosed that "[w]e shot all of the master shots in open water on the 1867 wreck of the *Rhone* in the British Virgin Islands and then went to the underwater set in Bermuda [to shoot the action]. We laced the set with live eels and live fish so as the camera panned around, you really felt that you were there."[56] Biologists named in the credits oversaw the selection and care of the fish used on the underwater set. Giddings recalled that when he saw the wreck set, it realized the shipwreck of his fantasies: "it was absolutely perfect."[57] To capture dramatic shark footage in the scene where the treasure hunters escape from sharks attracted by Cloche, the crew went to Australia. Further adding to the underwater impact was that Nolte, Bisset, and Shaw dove, in contrast to most underwater films, where the talent is doubled. "Fully 40% of the film takes place underwater and the actors and crew learned how to dive, playing long scenes without dialog on the ocean floor" noted *Variety*.[58]

The Deep played out its plot of adventure wreck-diving for treasure drawing on the great variety of atmospheres we have seen created on the underwater film set. The amateur divers are glamorous young beach bums hearkening back to the dive lifestyle showcased first in the Hass films, and we first meet them amidst the beauty of tropical waters. The action begins with a *Jaws*-style horror moment, combining a shot/countershot of predator/prey points of view with a jump scare. The horror effect occurs when Gail puts her hand into the crevices of a wreck, feeling for what turns out to be morphine vial, and it is seized by an unknown something that violently pulls her against the wreckage (it turns out to be a moray eel) (fig. 3.12). Adventure problem solving begins here too, as Gail pulls out her regulator, and releases its bubbles in the hopes that Dave will see the telltale signs of her presence and find her, which he does

FIG. 3.12. Peter Yates, *The Deep*, ca. 7:08. Gail pinned by an unseen monster.

FIG. 3.13. Peter Yates, *The Deep*, ca. 7:46. Dive craft: Gail using a trail of bubbles from her regulator to alert Dave.

(fig. 3.13). Such a *Sea Hunt*–style use of dive technologies to jury-rig escapes continues to the thrilling penultimate underwater sequence with the sharks filmed in Australia. In this escape, the treasure hunters are able to dodge the sharks by staying within an air column created by the dredge that Treece was using to move sand off the bottom of the wreck. The escape combines adventure thrills with the menace of shark carnage, which the film plays with rather than actualizing. Abstract clouds of blood come first, created by the chum that Cloche's crew use to lure sharks to attack the treasure hunters, while the white now comes from the air column enabling their escape, amidst the swirling sharks (fig. 3.14). Other moods evoked by the film undersea include the gothic

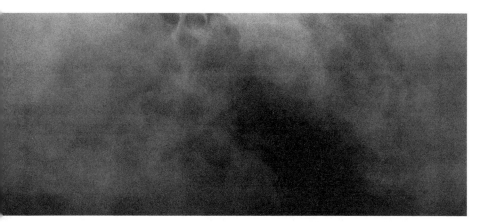

FIG. 3.14. Peter Yates, *The Deep*, ca. 1:34:52. The horror of shark carnage intimated by abstract clouds of blood.

FIG. 3.15. Peter Yates, *The Deep*, ca. 47:29. The monster as it strikes: a moray eel in overexposed black and white, as if captured by a still camera.

eeriness of the sunken *Rhone* when it is first explored by treasure hunters and a slightly surrealist horror sequence where Dave is attacked by the moray eel that is finally shown in action (like the shark in *Jaws*, a mechanical creature, which the crew called Percy). Terrified, Dave drops his camera that starts to trigger automatically, as its disco-like flash adds overexposed black-and-white imagery of the moray eel to the shot/counter shot pattern of predator and prey joined with disorienting jump cuts (fig. 3.15).

The Deep enjoyed box office success when it was released in the summer of 1977. Nancy Griffin and Kim Masters peg it as "the second-highest-grossing film for 1977 behind *Star Wars*."[59] Paul Moody records *The Deep* as "the tenth highest-grossing film at the US box office that year."[60] Underwater connoisseurs appreciated the film. *Diver*, the magazine of PADI (Professional Association of Diving Instructors) later called *The Deep* "one of the most loved and classic dive movies" of all time.[61] However, the film received tepid reviews from a more general public. *Variety* called it "efficient but rather colorless," despite its technical skill.[62] "'The Deep,' a Movie, Is Shallow" was the title of the dismissive review of the film in the *New York Times*, which praised the "beautiful" and "chilling" underwater scenes but dismissed the plot as "juvenile."[63] *TV Guide* judged the film "long, confused, and boring," with "dull undersea adventure that tried to capitalize on the success of the Benchley-Spielberg box-office smash *Jaws* (1975)."[64] The reviewer also called out the racism of a film where all the villains are Black. In his "scathing" write-up for *New York*, John Simon noted snidely that Bisset's appearance in a wet T-shirt without a bra might have "helped entice viewers": "the first image on the screen was of Jacqueline Bisset's breasts—a 'magnificently matched pair of collector's items.'"[65]

Guber had imagined "golden treasure gleaming in underwater sand drifts."[66] The filmmakers of *The Deep* were masters of creating captivating underwater effects. They were working with adventurous, charismatic actors. Why did *The Deep* lack imaginative traction for all but underwater connoisseurs? Perhaps its liquid fantasy was paradoxically diluted by its narrative complexity: two wrecks, differing techniques of diving and salvage, a range of villains and adventurers, and even perhaps by the film's skill at creating a variety of underwater moods. Such an answer, at least, is suggested by its contrast with the success of the monomaniacally focused *Jaws*, that repeated the shark attack over and over again to a completely fantastical climax. The next blockbuster in the lineage of underwater film after *Jaws*, James Cameron's *The Abyss* (1989), would continue the dive literacy evinced by *The Deep*, as well as its interest in the underwater film set as a theater for varied human action. *The Abyss* amplified these elements with the grandeur of *Jaws*, creating a liquid fantasy in the idiom of science fiction.

3.5. The Alien Depths Look Back: *The Abyss*

Throughout the 1950s and the 1960s, the undersea and outer space were for general publics and for scientists alike the two great cosmic frontiers. Yet, despite ongoing research, the oceans' depths lost their primacy in the public imagination by the end of the 1960s. Rozwadowski observes, "The cultural promise offered by the infinite extent of space came to resonate more strongly, by the end of the 1960s, than the fading dreams of fabulous wealth to be derived from the ocean's depths."[67] Puzzling on the turn away from the blue areas of the planet, Rozwadowski suggests, "The value of the ocean frontier rested in the vastness of its potential economic resources." When the ocean did not offer its treasure so easily, publics and governmental agencies seem to have abandoned its difficulties and looked into the yet more remote heavens.

Science fiction cinema from the late 1960s onward bears witness to the ocean's ebb as a site for its fantastic creation. The mid-1950s had seen fantasy beasts and monsters in films starting with *Creature from the Black Lagoon* and its sequels, including B pictures such as Roger Corman's *Attack of the Crab Monsters* (1957) and Arnold Laven's *The Monster That Challenged the World* (1957), with a killer "reptilian earthshaking beast of the sea" unleashed by an earthquake from the seafloor.[68] The television series *Voyage to the Bottom of the Sea* (1964–68) imagined otherworldly inhabitants of the depths; however, by the later 1960s, science fiction TV shows and movies would look to "space, the final frontier," to cite a phrase from *Star Trek*, whose original series ran from 1966 to 1969. In film, Stanley Kubrick's *2001: A Space Odyssey* (1968) and the *Star Wars* franchise begun by George Lucas in 1977 transported audience imaginations into outer space. Space continues to prime the science fiction imagination into the twenty-first century, with scores of B films, and also ambitious artistic cinema, including recently Alfonso Cuarón's *Gravity* (2013), Christopher Nolan's *Interstellar* (2014), and Denis Villeneuve's *Arrival* (2016).

Amidst our orientation to outer space, *The Abyss* (1989) is a blockbuster exception, pursuing the fantasy that aliens live in the abyssal depths. The director, James Cameron, credits his vivid sense of the undersea as cosmic frontier to growing up with *The Undersea World of Jacques Cousteau*, although Cameron lived "hundreds of miles from the nearest ocean." Thanks to Cousteau, Cameron, in his own words, "began to the think of the deep ocean as equal to outer space. . . . This was an alien world I could actually reach."[69] Cameron made his first dive outside a pool in 1971 and has maintained a career-long interest in ocean practice and in the aquatic imagination. Indeed, his ability to plumb the rich fantasy resources of the ocean despite the ebb of general cultural interest is one among his many directorial claims to fame, in *The Abyss* as well as in *Titanic*.

The Abyss created science fiction out of one of the least-known zones of the ocean, which it turned into a setting for adventure. The abyssal depths further

lent themselves to science fiction fantasy because they were so difficult to ac-
cess, beyond the reach of leisure divers, photographers, and filmmakers.
Hence, they were not robustly contoured in the popular imagination. Cam-
eron cites the image of diver/filmmaker as modern Prometheus from *The
Silent World*, at a moment in *The Abyss* when the protagonist, Bud, played by
Ed Harris, descends below seventeen thousand feet, to the limits of the ROV
Little Geek's ability to withstand the pressure. As he plunges even deeper, he
ignites a flare, taking the light of human invention to a realm unexhibited by
Cousteau and Malle, at once located far below the depths shown in their film
and also exuberant fantasy.

Cameron recruited as underwater director of photography Giddings, who
had a keen understanding of the difficulties of shooting on location. Instead
of photographing the planet's abyssal depths, as Cameron would later do at
the opening to *Titanic*, or simulating them in shallow water, Cameron adapted
unfinished tanks started for a nuclear power plant in South Carolina to create
"the largest underwater set ever built." This "feat of industrial engineering"
involved "pouring thousands of yards of structural concrete and installing
enormous filtration systems and pumps and a row of twenty-thousand-Btu
heaters to warm the 7.5 million gallon . . . to a comfortable temperature" for
directors and crew to spend entire days in there.[70] In these tanks, Cameron
would create a studio space to imbue submarine depths with the charisma
reserved for outer space in science fiction films from the time of *2001*.

Like previous filmmakers expanding the reach of underwater film and TV,
Cameron and Giddings oversaw new inventions to satisfy both the contours
of the film's fantasy and the demands made by its underwater film set. In the
chapter on *The Abyss* in *The Futurist*, Rachel Keegan surveys engineers' innovative solutions to the challenges of making the film, ranging from reconstituting
the ocean inside the plant's tanks to creating the aliens with a character generator. The tanks had to be specially lit, in order to create the color of the abyssal
depths, as well as filmed in a way that would not reveal their edges. According
to Charles Nicklin, the camera created the illusion of plunging into great
depths by traveling around the circumference of the reactor, instead of descending vertically.[71] One invention that won for its designers Richard Mula
and Pete Romano a special Academy of Motion Pictures Arts & Sciences
Technical Achievement Award in 1990 was "high-powered, safe underwater
set lights," the SeaPar 1200w underwater HMI lamp.[72]

Given the great depths where the story was set, the characters further
needed dive gear that did not—and still does not—exist. Cameron used this

opportunity to overcome an artistic challenge to drama from the inception of underwater film: the practical devices enabling divers to breathe and see, whether the dive helmet or regulator and mask, inhibit the actors' dramatic expression. Cameron requested from designer Ron Cobb a "dive helmet with a clear faceplate, so the actors' every subtle facial movement would show."[73] Further, he asked Western Space and Marine, the company building the helmet, to incorporate "the regulator into the side of the helmet . . . cycling fresh air as needed without a mouthpiece, freeing the actors to talk. To record the dialogue, microphones normally used in fighter aircraft helmets were incorporated into the WSM helmets."[74] Cameron's interest in enabling expression extended to the aliens. Since "traditional visual effects methods proved inadequate," Cameron worked with "VFX house Industrial Light & Magic to create the 'water tentacle,'" the aliens' vessels to contact humans.[75]

With such technologies Cameron innovated not only a new frontier for science fiction but also a new cluster of underwater science fiction effects, distinct from the aesthetics of outer space in science fiction film. An influential example of outer space creation for comparison is Stanley Kubrick's *2001: A Space Odyssey* (1968), which Cameron remembers having had "a strong physiological response to," when he first saw it.[76] As Kubrick constructs outer space in this film, it is an awe-inspiring place that appeals to the sense of sight through offering revelation. Beyond the light of this sun, other stars and galaxies gleam and glitter in the inky void that promise new worlds as far as the eye can see. This light is cold, exciting our intelligence as the path to contact with other alien realms. While such light burns, the fire is too far away for danger—instead it is just bright enough to see, if not to feel. Its light intrigues with its differing intensities and patterns, as well as with the movement of appearing and disappearing astral bodies.

In this cosmic vacuum, bodies have a satisfying, three-dimensional tangibility and mass. These futuristic projections appeal to the classical taste for rounded, symmetrical forms, in keeping with the intellectual aura of space representation. Kubrick's imagined outer space in *2001* was shaped by pictures of this realm drafted by Chesley Bonestell, one of the most influential creators of outer space fantasy in the twentieth century. Bonestell came from a background in architecture and brought to space an architect's love of geometry. Bonestell's imagery would influence Clarke, author of *2001*, who indeed provided text to a book of Bonestell's illustrations. "According to *2001* special effects supervisor Douglas Trumbull, Bonestell . . . [also had] a tremendous influence on director Stanley Kubrick," an influence that is visible in Kubrick's rotating space station and throughout the astral imagery of the film (fig. 3.16).[77]

FIG. 3.16. Stanley Kubrick, *2001: A Space Odyssey*, ca. 1:31:40. The vast visibility and sparkling contours of outer space fantasy.

The aquatic atmosphere thwarts such a satisfying visual sense of deep space, mass, and tangibility. While the underwater haze can be relaxing at the surface, as a respite from the brightness of the sun, and while we have seen the haze add glamour to images of the body, haze becomes gloomy the moment the plot turns menacing. Sublime effects, as Enlightenment philosopher Edmund Burke knew, are produced from obscurity as well as from confusion. Thus, Burke cites as "dark, uncertain, confused, terrible, and sublime to the last degree" Milton's description of death in *Paradise Lost*. "The other shape, / If shape it might be called that shape had none / Distinguishable, in member, joint, or limb; / Or substance might be called that shadow seemed."[78] The underwater haze lends itself to such menacing sublime effects, particularly when melded with a confusing plot and disruptive camerawork and shots. Underwater haze works against the Olympian promise crystallized in the reach of unlimited visibility that belongs to the fantasy of science fiction, even when everything is going wrong.

Thus, contrast the atmosphere, both emotional and physical, when astronaut Frank Poole, played by Gary Lockwood, spins off into space with his air hose cut, to that saturating the many disasters that afflict the *Deepcore* rig. Kubrick creates a stunning, sublime discomfort from juxtaposing the beauty of the heavens, which spectators still can master with their gaze, with the character's chilling fate.[79] Many of the *Deepcore* disasters, in contrast, are shrouded in disorienting fog. A good example is the scene where Lindsey Brigman, played by Mary Elizabeth Mastrantonio, tries to gain more oxygen for the *Deepcore* by hooking it up to storage tanks outside the partially destroyed rig on the edge of the abyss. Lindsey moves slowly through the resistant water, in contrast to the absence of friction that gives portrayals of humans floating in outer space a remarkable ease. As Lindsey works methodically but with extreme urgency, both her figure and the machines are blurry in ominous gloom.

The ominous mood Cameron creates from underwater optics and the depths' impact on the body enhances Lindsey's—and the spectators'—marvel at the violet, glowing NTIs (non-terrestrial intelligences) when they appear in this scene. As Lindsey's visor reveals her astonished face, the expressive helmet Cameron introduced to let actors emote underwater pays off. Lindsey's pose is expressive as well, with arms outstretched to touch the alien craft, echoing the marvel of the spectator at an aquarium. The apparition that inspires her awe synthesizes multiple optical effects that have previously created wonder underwater. Along with the softly luminous colors in the haze, the slowed motion turns from gloom to beauty, reprising the underwater dance. As Lindsey tells her crew on the *Deepcore* about her first meeting with NTIs, "It glided. It was the most beautiful thing I've ever seen. . . . It was a machine but it was alive. Like a dance of light" (ca. 59:20).

In making the film, Cameron was inspired by H. G. Wells's "In the Abyss" (1896), which recounts an explorer's descent in a deep-sea submersible to a ghostly, repugnant realm. Still shaped by the pre–helmet diving vision of the undersea as a haunted Davy Jones's locker, littered with corpses of the drowned, Wells imagined this realm as inhabited by reptilian creatures, whose houses have structures resembling bones, appearing "as if [they] were built of drowned moonshine."[80] In contrast, Cameron's NTIs are "beautiful yet insubstantial, comprised of sheer, watery light," in the words of Stacy Alaimo.[81] Alaimo connects Cameron's imagination of the abyssal depths to the history of deep-sea exploration, starting with William Beebe's account from the bathysphere of abyssal bioluminescence. For Beebe, Alaimo writes, "[t]he enthralling blue-black light, as well as the colors emanating from bioluminescent creatures," in these depths "destabilize terrestrial assumptions, values, and modes of life," as she identifies "a nascent posthumanist philosophy."[82] In the case of Cameron, in contrast, Alaimo challenges his commercialism, and his replacement of deep-sea reality with myth. She contrasts Cameron's vision with Claire Nouvian's documentary book of photographs, *The Deep: The Extraordinary Creatures of the Abyss* (2007). These photos, in Alaimo's words, are of a "violet-black abyssal ecology [that] is not a single hue but is instead populated with a spectrum of colors," inhabited by "[p]rismatic, fluid constellations of bioluminescent animals."[83] At the same time, Nouvian's book recalls Cameron's science fiction fantasy of deep-sea beauty from its title to its imagery, suggesting that blockbuster liquid fantasy is not so easily untangled from less commercial art.

In *The Abyss*, the ability to experience marvel at the NTIs is only available to those characters with water knowledge so deep they can draw on it for lifesaving intuitions. We might expect this knowledge from the Navy SEAL Hiram Coffey, played by Michael Biehn, but he is overly dependent on terrestrial technologies and goes crazy from a science fiction form of rapture of the deep. Bud, in contrast, is in tune with his environment, and his commonsense problem solving is shown, for example, when he uses a sailor's knot to restrain Big Geek, which Coffey has commandeered, in the ultimate submersible battle with him. Lindsey, too, possesses such knowledge, as when she entrusts her life to humans' evolutionary physiological connection to the sea, the mammalian dive reflex (MDR) used by free divers that slows down body functions. Giving herself over to this reflex, she drowns, with the potential for resuscitation in an adventure sequence where she and Bud have only one set of the film's futuristic diving gear to make it back to the *Deepcore*. The science fiction prostheses being essayed for deep diving without a submersible in this film also point humans toward aquatic existence, following Cousteau. The film imagines breathable liquid, which would enable people to spend time in great depths without their bodies and lungs being crushed.

The culminating appreciation of the undersea offered by the film is its invitation for spectators to experience the NTIs' point of view. We have seen the remarkable, if brief, sequence giving us the dolphin-eye perspective in *Flipper*. Twenty-six years later, Cameron similarly creates a liquid, distorted wide-angle shot expressing the underwater eye that he associates with the perspective of the NTIs. The film introduces the perspective of the NTIs when an alien probe enters the rig through the pool that connects the *Deepcore* to the sea (fig. 3.17). In this sequence, the camera shows us how the NTIs see the rig, estranging terrestrial atmosphere and forms, which are fluid, stretched, and wavering. In contrast to the cold hostility of Kubrick's malevolent AI, HAL, the shimmering gaze of the NTIs endows humans and even their technologies with luminescent beauty.

FIG. 3.17. James Cameron, *The Abyss*, ca. 1:07:32. The liquid perspective of an NTI.

Various critics have noted "Cameron's investment" in *The Abyss*, as Alexandra Keller observes, "in multilayered visual technologies . . . such that the spectator never goes very long without having to negotiate such intervention."[84] These diverse visual technologies draw attention to the produced nature of film spectatorship and to the tenuous boundary between technology and the human. Yet their presence also is useful, in a basic, practical way, in the atmosphere of water, where we need prostheses for breathing and seeing. Underwater, mediated vision expands human access and experience. Indeed, in *The Abyss*, Cameron emphasizes the difficulties humans have seeing underwater that land-oriented spectators might not appreciate. At one point, the survival of the rig's crew hangs on Bud's ability to distinguish a blue wire with a white stripe (the ground wire he must cut to disarm a nuclear warhead) from a yellow wire with a black stripe (the lead wire that can blow them up). The engineers writing Bud's instructions on land did not take into account that such color nuances are not perceptible in the depths of the sea, even with artificial light that bathes everything in a yellowish glare.

In the idiom of popular cinema, Cameron is extending the invitation to look with a marine eye that Man Ray offered in the avant-garde *The Starfish*. While Man Ray's interest in marine vision leads to surreality, Cameron strives for "a biocentric" perspective, including the marine environment.[85] Lindsey declares that people see what they want, and that "you have to look with better eyes than that" (ca. 1:00:13). Her better eyes acknowledge the aquatic areas of our planet and humans' sustaining relation to it. Only with these better eyes does Bud descend to the abyss, where, when he finally reaches bottom, he is saved by the NTIs. In this shot/countershot montage of meeting, Bud's gaze and the NTI's gaze are juxtaposed, suggesting meaningful interspecies exchange, reinforced when the NTI reaches out its hand, which Bud takes. Then, Bud is treated to marvelous sights: the aliens' realm inspired by those aspects of undersea life that have most fascinated scientists and publics, such as the qualities of form, color, and bioluminescence. The film foreshadows Bud's ability to access this marine vision as he leaves on the mission that will plunge him down to the alien's home. Before diving in, he takes one look back at the group on the *Deepcore* through the portal pool and sees his familiar crewmates with an estranged aquatic view, undulating through water.

Keegan points out that Cameron's "script started with a quote from Friedrich Nietzsche: 'And if you gaze for long into an abyss, the abyss gazes also into you.'"[86] While Cameron dropped this quotation because it had recently appeared in another movie, it suggests a reading of the film, in Keegan's interpretation, where characters "confront a monster, and find out the monster was them."[87] She emphasizes human self-absorption, finding a likeness in the abyss that we relate to ourselves, and yet the word "gaze" is equally important. The NTI gaze turns out not to be a mirror of bottomless existential dread, as might be suggested by the decontextualized citation from Nietzsche, but rather a more benevolent, interspecies curiosity. Furthermore, Cameron's NTIs use their ability to control water to simulate the human countenance, gazing back at humans to ask them urgent questions about planetary exploitation and despoliation.

Cameron's interest in the abyssal depths as a frontier for vision, aesthetic as well as scientific, may have shaped his own subsequent actual adventure in a probe down to the deepest place in the ocean, the Mariana trench, which he achieved in 2012. In a film made with *National Geographic*, *Deepsea Challenge 3D* (2014), Cameron realizes a version of the fantasy dive that Bud was forced to take only in his suit. At the bottom, Cameron's first glimpse was of a barren floor, a "lunar," "alien" world.[88] Although Cameron filmed this actual "alien" environment, its life forms do not enrapture, like those created by Industrial

Light & Magic. Fact in these images is astounding, yet far less immediately charismatic than liquid fantasy.

The money Cameron spent on his probe and the subsequent film returns us to Alaimo's critique of celebrity adventurers, from Beebe to Cameron, in a lineage also including Cousteau. Do adventurer-exhibitors help give people a sense of ocean reality and hence the need for ocean research and knowledge, or does their entertainment substitute for it? The question does not have a simple answer. Such people produce best-selling imagery and inspire general interest in the undersea—a goal of environmentalist activists and scientists alike. At the same time, their expenditure does not contribute to the ongoing needs of the more prosaic, professional marine scientists competing for resources in underfunded fields of research. In the words of Oxford University biologist Alex Rogers, quoted in the *Guardian* in 2016, "Deep sea research needs to be funded at a similar scale to space research. It's as simple as that. We want to get more people engaged in the deep seas, to feel inspired and care more about them. Hopefully, people will then start to demand they are managed properly."[89]

3.6. The Blue Mind: *The Big Blue*

The Big Blue fictionalizes the lives of two men who shaped the modern sport of free diving: the Frenchman Jacques Mayol, who was the first person to reach the depth of one hundred meters in 1976, and one of his rivals, the Italian Enzo Maiorca, renamed Enzo Molinari in the film. Luc Besson, like Cameron, is a filmmaker steeped in the practice of the water from childhood, since his parents were scuba instructors for Club Med. When Besson recalls first learning about Mayol by watching a documentary on the free diver when he was sixteen, he emphasizes the visual as well as human impact of Mayol's exploits. In Besson's words, "It was not the fact that he descended into blackness, without breathing for four minutes, subject to enormous pressure that made me cry," it was rather his smile when he came up after such an incomprehensible, terrifying journey, and how relaxed he looked.[90]

Besson by his own account had to stop diving because of an accident whose details he has never shared. Nonetheless, according to the credits of *The Big Blue*, he did some of the underwater camerawork for the film, assisted by the noted underwater photographer and cinematographer Christian Pétron. Pétron came to *The Big Blue* from documentary photography, where he has continued his illustrious career, shooting as technical director for the Discovery Channel for films about the exploration of the *Titanic*, and filming sharks and whales with Jean-Michel Cousteau, Jacques Cousteau's eldest son.

When the film premiered at the Cannes Film Festival in May 1988, in Susan Hayward's words, "it was pilloried by critics and the audience that saw it. The music was too invasive, the film too slow, the characters lacking depth. . . . Why open the festival with an English version of an all-French film subtitled in French?"[91] The film likewise was received quizzically by general audiences in France and the United States. Thus, in the *New York Times*, Janet Maslin wrote, "Although 'The Big Blue' was made in English by a French director, it does not show signs of being workable in any tongue, save perhaps for the language of the briny deep. And a lot of what happens here would beggar even a dolphin's understanding. But the film has a handsome, expensive look and a charmingly inscrutable manner that make it easier to take than it would have been staged more solemnly. The scenery is glorious, and that doesn't hurt either." The picture accompanying the review is a topside photo of Rosanna Arquette, rather than one taken from the film's many underwater scenes.[92]

The Big Blue would go on to success with general audiences, who appreciated its existential engagement with the depths and the elusive sea. At the same time, *The Big Blue* is the work in Besson's oeuvre least discussed by critics.[93] Hayward, one of the few to take it seriously, reads it as directing a strong ecological message against corporate "greed and graft."[94] Besson would go on to elaborate this message in his subsequent film, also made with Pétron, *Atlantis*, a documentary featuring the marine environment as a place of strangeness and great beauty.

The Big Blue's existential preoccupations come into focus with the context of liquid fantasy and the history of underwater film. The film constructs memorable imagery evoking the unrealizable human longing to penetrate the barrier of the surface and join air and water in one harmonious existence. Besson took up this human attraction to an element that was at once life-enhancing and destructive in his first screenplay for a short, *The Little Mermaid* (*La p'tite sirène*). This film "tells the story of a free-diver who eventually lets go of the plumb line and goes off into the deep blue with a siren."[95] In *The Big Blue*, the ambivalent relationship to the deep blue is the core of the drama that grips its hero, Jacques Mayol. The film does not hold us in suspense about Jacques's record-breaking dives but rather about the tension between his attraction to the water and his ability to function on land. Free diving strips the body of its reliance on technology, or perhaps uses the body as the only technology, removing the pervasive underwater theme of technological innovation from the characters' underwater adventures. In Besson's account, the challenge of the human relation to water is primarily psychological and existential, even if technology aids its realization.

FIG. 3.18. Luc Besson, *The Big Blue*, ca. 36:18. Dolphins as Jacques's requited yet distant love objects.

Dolphins are the mammals that embody Jacques's desire to merge with the water. "Don't think of Jacques as a human being. He's from another world," says Enzo, played by Jean Reno, to Johana, the character played by Rosanna Arquette, in love with Jacques (ca. 54:19). The film abounds in scenes of dolphins communicating and cavorting with Jacques, in the sea, in oceanariums where Jacques works, and in Jacques's dreams. This connection recalls the anthropomorphizing vision of dolphins chronicled by Mitman. It also evokes the film's real-life model. Mayol authored *Homo delphinus* (1983), in which he wrote about his life as a free diver and his fantasy of human kinship with dolphins.

Throughout the film, Jacques demonstrates a mystical bond with dolphins. These animals share his emotions, whether he wins a competition and dolphins in oceans and oceanariums jump in joy, or a dolphin comes to greet him by the bright disk of the full moon, cleaving the shimmering surface of the sea with exuberant somersaults, after he and Johana have sex. This connection between man and dolphin has matured from the youthful friendship shown in *Flipper*, but it is a variation on a theme. From Jacques's first glimpse of the dolphins as a child to the closing image, the film reinforces the dolphins' presence as intermediary figures in a modern mythology of the elusive, irresistible, unconsummatable love of humans for the water; mediating between human and the sea, between life and death—and also between reality and dream (fig. 3.18).

In tension with Jacques's attraction to the water are calls from the land, offered through a bromance with his fellow free diver Enzo and a romance

FIG. 3.19. Luc Besson, *The Big Blue*, ca. 30:17. Free diver as superhero.

with Johana. Neither shares his relation to the sea. For Enzo, the depths are the arena where he can display mastery and thus solidify his status on land. Johana desires roots—a baby, a home, a car, and a dog, as she tells Jacques. From their first meeting, the film shows Rosanna repeatedly unable to connect with Jacques. This meeting is high up in the Andes, where Jacques is diving in a frozen lake. A doctor tracking the dive explains the mammalian dive reflex that enables free diving and Jacques's extraordinary underwater adaptation. While Johana watches Jacques's dive from the trailer, she can see only his remarkably slow heartbeat. She cannot see the stunning views of Jacques feeling his way along a crust of ice in a freezing lake that we are shown thanks to an underwater camera. In this feat, the ice and aqua blue water, glassily silver, contrast with Jacques's Spiderman-like superhero figure in a red wet suit (fig. 3.19). This split between Johana at the surface and Jacques in the depths has a sad reverberation in a later scene when Johana tells Jacques she is pregnant. To get his attention, she jumps off a cliff into the sea—only to have him follow her but then plunge below, leaving her alone, screaming in frustration.

Laurent Juillier observes that "many people commented on the fact that [the] film's undersea shots are no better than the most ordinary sequences of Jacques Cousteau's cinema."[96] From a practical standpoint, Besson did not have at his disposal the budget or underwater production resources of Cousteau—or of Cameron. From an aesthetic standpoint, Besson strips away a complex use of technology from the achievements of the film, just as he has stripped away the fascinations of underwater technology from the plot. The framing of shots is

FIG. 3.20. Luc Besson, *The Big Blue*, ca. 1:21:26. The human body: the free diver's ultimate technology.

simple, to highlight the human body underwater. An example of such simple framing is a view of Jacques at the bottom of his record-setting free dive, being checked by scuba divers. While in a Bond film, their scuba gear might have been displayed as hypermasculine, in *The Big Blue* these characters seem technologically burdened and fade into the setting. The camera instead lingers on Jacques's calm underwater, and among the free divers, Jacques is unique in eschewing even the simple prostheses of goggles (fig. 3.20). In this image, he looks at the scuba divers with a beatific countenance, recalling Besson's own account of the impression Mayol made on him by finishing his superhuman achievements with a relaxed smile.

Has Jacques achieved the ability to see underwater perhaps mercifully denied to him in the childhood trauma when his father drowned? In this scene, the child strains to catch a glimpse of his dying father and cannot—whether because he is being held back or, when the screen fills with bubbles, because the child does not have the layer of air his eye needs for such vision. The scene of a sponge diver drowning in the presence of his son, who then dedicates himself to diving, may be a nod to Robert Webb's *Beneath the 12-Mile Reef*. If so, it is one of several allusions to the views from the history of underwater film in the opening sequence from Jacques's childhood, shot in black and white to place it in the past. Thus, the hero Jacques Mayol's opening dive as a boy is shown first from the depths, from a point of view looking up that is not associated with a character. This shot sets up an expectation of vulnerability

FIG. 3.21. Luc Besson, *The Big Blue*, ca. 3:49 (opening in black and white): Jacques the boy and his friend the moray eel.

for the reed-thin boy who, however, proves so at home in water that such vulnerability, here, as throughout the film, turns out to be only a teaser. Further teasing us in this scene is the boy's object, which is to feed a moray eel. Jacques goes straight to the crevice where the creature is hiding, suggesting that it is an old friend, in contrast to the moray eel in *The Deep* (fig. 3.21). Perhaps yet one more allusion in the initial sequence where Jacques is feeding the eel is a shot from a perspective inspecting his backside. The creature associated with this point of view proves to be a dolphin, whom the film shows mimicking Jacques's movements. Did Besson know about the hypersexuality of dolphins that was a challenge for oceanariums trying to integrate dolphins into a general audience family narrative? As Mitman details, for example, Flippy, the 1950s performer at the origin of Flipper, "was himself known among Marine Studios personnel for his ceaseless 'masturbatory practices.' Often, the sexual aggressiveness of dominant males resulted in the death of other dolphins and specimens in the tank."[97] In any case, Jacques's first view of a dolphin causes him to fly to the surface in a panic, although, the film then cuts to him waking up inhaling in a gasp, confusing the spectator about whether the encounter occurred in the real-world narrative of his film or in his dreams.

As an adult, Jacques puts himself in sync with the depths by pursuing the sport of free diving. Besson creates a distinctive point of view to show us Jacques's mindset as he prepares for the first competition shown on screen. Free divers prepare with a series of breathing exercises that saturate their blood with oxygen. A state of extreme relaxation accompanies this physical preparation. When it is Jacques's turn to dive, he emerges on deck in this state that might now be called the blue mind, in reference to a book of that name by Wallace Nichols. In a 2017 interview with *USA Today*, Nichols summarizes this notion as "the mildly meditative state we fall into when near, in, or under water. It's the antidote to what we refer to as 'red mind,' which is the anxious, over-connected and over-stimulated state that defines the new normal of modern life."[98] A slow, rocking motion characterizes Jacques's blue-mind point-of-view, as if he were already fluid as water, set apart from those around him, whose gestures appear at once well-meaning but irrelevant. In this scene, we can also grasp how Jacques's blue mind transmits to others an affect of melancholia and disconnection. The character Jacques has been read as traumatized from his father's drowning, and yet he could also be considered to live life on land in a perpetual state of the mammalian dive reflex, a state that sex and love do not affect, broken only with Enzo's drowning.

Pool sequences, as I elaborate in the next section, are a way for films to question the values of life on land. Besson's version, suggestive of surrealist *piscinéma*, takes place in the swimming pool of a Taormina hotel, where Enzo and Jacques drink champagne underwater, which, following a toast, turns into a drunken breath-holding competition. The visuals are stunning displacement (*dépaysement*) of social pretension. Jacques and Enzo sit in elegant tuxedos, using the bases of outdoor furniture to weight themselves down as they simulate drinking champagne underwater, toasting and even looking at the label of the bottle as if admiring the vintage. Their glasses submerged in water turn inside out the relation of the container—the pool and glass—to what is contained—water and champagne—and by implication, social relations (fig. 3.22). Beyond mocking the rituals of the wealthy, this beautiful scene expresses the impossible camaraderie of Enzo and Jacques, at first exuberant, but also grimly determined, as difficult as drinking champagne underwater, and as temporary as the time they can spend there, which moreover ends in a blackout.

The blackout ending this scene returns amplified in the despairing blackout at the film's close. Its preparation is Jacques's final dream sequence, after he has been injured in an inconsolable, unregulated dive following Enzo's death. Sick in his hotel bed at night, he has a vision that the depths flood down from the ceiling into his bedroom. In this vision, the water's surface descends slowly with the surface flipped, as if seen from underwater, giving a hallucinatory feel to the imagery, which engulfs him with its luminous dance of aquatic light (fig. 3.23). Once Jacques is submerged, the camera performs a dramatic 180-degree rotation, as if expressing the easy mobility possible in the depths.

FIG. 3.22. Luc Besson, *The Big Blue*, ca. 59:09. Social pretensions displaced: Jacques and Enzo drinking *in* the pool.

FIG. 3.23. Luc Besson, *The Big Blue*, ca. 2:34:48. Jacques dreaming of impossible human melding with the sea.

At its conclusion, the water's surface viewed from beneath is restored to its expected orientation. However, the sequence introduced with this camera movement is unnerving. The spectators see dolphins cavorting in a beautiful but placeless blue atmosphere—the big blue—without a human in the frame.

Following this vision, Jacques undertakes a suicidal nighttime free dive. Although the real Jacques Mayol lived to the age of seventy-four, the character's plunge seals the unbridgeable chasm, at least for humans, between terrestrial and aquatic life. Jacques is tormented over his rival's death, his girlfriend bearing a promise of new life, and his own longing for the blue world he can never completely join. Imbued with hope and desire in childhood—in keeping with the echoes of the children's film *Flipper*—Jacques's longing has now turned to madness. In the film's final sequence, Jacques ecstatically reaches toward the snout of a dolphin in the dark, velvety depths, before the film gives a black screen, the conventional color of the screen at a film's ending coinciding with the trained diver's blackout at the limit of oxygen, at which point he or she will drown if unassisted, as Jacques is in this scene.

3.7. Aquatic Perspective Revealing Warped Society: From *Sunset Boulevard* to *Inception*

The swimming pool *dépaysement* in *The Big Blue* sits in a longer tradition of pools exposing warped reality on land. Since Man Ray's *piscinéma* in *The Starfish*, the view of land life from the pool has used the optics of water to suggest something amiss with everyday reality. One of the first instances of such perspective in Hollywood cinema is the famous shot of the dead body floating in the pool that opens Billy Wilder's *Sunset Boulevard* (1950), captured from, in Wilder's words "a fish's viewpoint."[99] "The shot proved to be so central to Wilder's vision of the film, and so difficult, that several whole days . . . were spent solving the problem of how to do it."[100] Lacking scuba gear, and apparently without the ability to film underwater, associate art director John Meehan came up with a way to achieve this shot inspired by "a magazine article on . . . the way fishermen look to the fish they are trying to catch."[101] This technique involved "the camera pointing down toward a mirror on the floor of the tank. Behind the policemen was a large piece of sky-colored muslin." In this arresting sequence, the swimming pool goes from an expression of social status and a gathering place of delight and relaxation to a witness, if not an accessory, to murder. The view from the pool is "spectacularly macabre," as it shows us the corpse from below, with behind him, through the water's surface, the police staring and the flash of a photographer's bulb.[102] "The audience is unnerved not only by the ghastliness of the corpse," whose face we see, to enhance the horror, "but also by the position we are asked to assume. . . . [W]e . . . have sunk to the bottom."[103]

In *Sunset Boulevard*, Hollywood celebrity is under scrutiny. In *The Graduate* (1967), Mike Nichols used this view from the bottom of the swimming pool to highlight the vapid materialism of affluent American suburbia. Robert Beuka contextualizes this scene in relation to the topos of the swimming pool

FIG. 3.24. Mike Nichols, *The Graduate*, ca. 22:39. Scuba is all the rage in 1967 Southern California, and Ben tries out his diving gear in his parents' Pasadena kitchen.

in literature and film of the mid-twentieth century, such as *The Great Gatsby* and John Cheever's short story "The Swimmer." In such works, the pool is a kind of mirage—it epitomizes material success and holds out the promise of leisure, but in fact, it reveals the "depths of despair—that lurk beneath the shining surfaces of suburban life."[104]

Nichols first introduces water into the suburbs with Benjamin's fish tank, which would plausibly have been kept by a middle-class boy in the 1960s, according to the timeline established by Judith Hamera.[105] At the same time, Ben's preoccupation with the aquarium signifies his distance from people around him; as Sam Kashner writes, "Benjamin sees the world through glass."[106] Ben is equally distanced from his society in the pool sequence, initially comic, when he then tries out scuba gear in his family's Pasadena home. We first see him standing in the kitchen, flippers facing forward, out-of-place (in contrast to the cool image of the diver over at the coast on the Southern California beaches at the time) (fig. 3.24). His comic appearance gives a random feel to the smug suburban kitchen, recalling Salvador Dalí's challenge to avant-garde art and intellectuals when Dalí appeared in a helmet-diving suit in the midst of the International Surrealist Exhibition of 1936 (where he would start to asphyxiate and have to be freed) (fig. 3.25).

Although *The Graduate* sets us up for realism, the mood turns surrealist in the pool sequence, culminating in Ben's dive, where he escapes from his family and looks back up at them in their suburban backyard, through the veil of water. The producer of the film, Lawrence Turman, identified this scene as one of the two that gripped him about the novel (along with the final scene of Ben and

FIG. 3.25. "Dali in diving suit and helmet with Paul and Nusch Eluard, E.L.T. Mesens and Diana Brinton Lee and Rupert Lee at the International Surrealist Exhibition June 1936." Dalí donned the diving suit for his lecture but soon had to be freed, as it became apparent he was suffocating. Photograph. National Galleries of Scotland. Purchased with the assistance of the National Heritage Memorial Fund and the Art Fund, 1995.

Elaine on the bus, Elaine in her wedding dress): "a boy in a scuba suit in his own swimming pool."[107] It was also one of the most difficult to film. Robert Surtees, the master cinematographer hired to make the film, told his operator and assistants, "You fellows be prepared because you're going to do some way-out shots." In a piece cited by Kashner, "Using the Camera Emotionally," Surtees recalls that "'we did more things in this picture than I ever did in one film. . . . We used the gamut of lenses . . . hidden cameras, pre-fogged film,' as well as hand-held cameras. In one particularly difficult shot that a special camera operator had to rehearse for two days, Surtees's camera acts as Ben, as he walks out of the house in wetsuit, diving mask, and flippers, dives into his parents' pool, swims underwater, and resurfaces, only to be pushed back into the pool by his father."[108]

This shot from Ben's point of view lasts about a minute and a half—an exceptionally long time in Hollywood cinema—and features a number of "way-out" elements besides its length. The shot uses an oval frame, blocking off the action and thus creating a sense of partial and frustrated viewing. This frustration is reinforced by the unstable, handheld camera, expressing how Ben staggers toward the pool in his flippers, under the weight of the tank. The comedy of maladaptation continues as the camera looks down at Ben's flipper tips intruding into the view frame, then rotates as it breaks the surface and dissolves the firmness of land into an aquatic point of view. The radical camera movements enabled by the underwater atmosphere—such as at least 270- or perhaps 360-degree pans around the sides of the pool—thwart orientation, given the swimming pool's uniformly colored sides and floor, undifferentiated except for a ladder and the play of light. The spectator's disorientation in the aqua depths sets up the subsequent dystopian view of Ben's parents, his mother overexcited and his father pushing his son back into the water. Seven years before *Jaws*, Surtees realized the power of a shot hovering at, above and just below waterline, which is indeed in this film, "a harrowing transition zone" (Elias).

With our eyes saturated by the aquatic point of view, we are primed to recognize the residue of its distortions in the twisted relationship between Ben and Mrs. Robinson. Our first view of the two meeting at the Taft Hotel bar is upside down in a mirrored table, as if reflected in a pool. The characters' forms are subtly blurred through the smoky atmosphere of the bar, reminiscent of the underwater haze. The association of the aquatic perspective with transgression continues as well in the sequence that follows their encounter, where we see Ben, sometimes lying down, sometimes floating in the pool, presumably daydreaming, with imagery that includes superimpositions of the undulating aqua atmosphere, punctuated by intermittent dazzles of light, so that spectators cannot at first tell whether the images are under water or in air. When Ben subsequently sees Mrs. Robinson, now accompanied by her husband and his parents, he is again floating in the pool on an air mattress, and she appears through a fog on his sunglasses (fig. 3.26).

In both the pool and seduction scenes, Nichols evokes the underwater film set to create a liquid fantasy without a connection to the subject of the sea. Rather, aquatic optics offer a way to express distorted social norms and the perversities of individual desires.[109] This liquid fantasy in *The Graduate* resonates with the concept of "enwaterment," used by Adriano D'Aloia, which he suggests as extensive notably in twenty-first-century films that "both physically and psychically engage . . . the spectator in a 'water-based relationship'"[110] In these films, in D'Aloia's words, "[c]inema literally and metaphorically seeks to construct a 'water-based' environment, a sharable site

FIG. 3.26. Mike Nichols, *The Graduate*, ca. 44:38. Mrs. Robinson through the fog of Ben's sunglasses. As in *L'Atalante*, the marine layer expresses terrestrial desire.

of experience . . . by the extension of the expressive properties of water outside the fictional space of the screen." D'Aloia points to a number of films, with varying moods, where a plunge beneath the surface or a suggestion of aquatic haze will be the transition to a world of fantasy, hallucination, and dream. With their swimming pool moments, comic films as well—like *Ferris Bueller's Day Off* (1986) and *Old School* (2003)—join a lineage inaugurated by avant-garde interest in underwater optics. D'Aloia also mentions a memorable moment in Spielberg's *AI* (2001), in an homage to *The Graduate*, when David, the robot, a "mecha" in the film's terms, is abandoned after he almost drowns his human "brother." Staring up at the surface of the pool from its bottom, we get the mecha's point of view, where "[t]he water surface acts as a sight-filter that offers a view into an altered, faraway and hostile world."[111]

Spielberg used underwater optics for sinister expression in his darker science fiction film, *Minority Report* (2002). There, the swimming pool is the location where the hero, Anderton, played by Tom Cruise, loses his son. Distorted shots of the surface, blurred action, and Cruise's perspective underwater in the pool block perception of the moment when his son disappears. Moreover, the aquatic haze infiltrates terrestrial scenes, as when the Department of Justice agent Danny Witwer, played by Colin Farrell, looks around Anderton's office and the camera reveals photos of his dead son, or the multiple views in the PreCrime headquarters in the scenes shot through glass, which here is used in instruments of coercion and unjust framing—of both vision and reality.

The minority report is the key that unlocks the dark secrets of this society where unfettered capitalism utilizes numbing surveillance to dominate and indeed enslave citizens. The report takes the form of enwatered imagery evident when the panicked psychic Agatha, played by Samantha Morton, reaches out from her pool and clutches Anderton, uttering "can you see?"(ca. 28:47). The rapid, blurry montage of terror that follows shows a woman being murdered and then submerged in water. Multiple hazy layers contribute to this blur: the mist from the precogs' pools, shots into and through the water of the countenance of the dead victim, along with the projection of Agatha's visions onto a screen. In contrast to the mesmerizing effect of *piscinéma* as used by the surrealists, *Minority Report*, like *The Graduate*, uses aquatic optics to express dark disturbance, both Anderton's tragic loss and Burgess's criminal murder of Lively.

Christopher Nolan's *Inception* (2010) is one film where aquatic optics appear in a science fiction plot of such intricacy that spectators lose the ability to identify what social values are being distorted, and are encouraged instead toward an existential reading. From the relaxed, impenetrable high of Jacques Mayol in his element in *The Big Blue* to the livid face of the drowned man in the swimming pool in *Sunset Boulevard* or the anguished victim submerged in *Minority Report*, human faces underwater are particularly commanding in their vulnerability. *Inception's* contribution to this range of aquatic physiognomies is an image of a human estranged from himself. In the crash scene of a van falling off a bridge, the face of the hero, Cobb, played by Leonardo DiCaprio, is impassive and unreadable, as he sways back and forth in the movement of the river. No oxygen escapes—he is presumably blacked out, or drowned, yet he looks relaxed, as if he were sleeping. Has Cobb descended to limbo, stuck forever in the timeless world of a dream, or will he find an escape? His enigmatic visage is in keeping with the film's ending, which depends on our reading of its final image of the wobbling top.

Reenchantment is one more possibility when the underwater film set is used to express aspects of society on land. Although not a popular success, M. Night Shyamalan's *Lady in the Water* (2006) features an arresting use of the aquatic haze and underwater optics to reenchant a downscale apartment complex in Philadelphia, appropriately named The Cove. The occasion for such effects is the appearance in the pool of a water nymph, which the film calls a "narf." The narf is one of the wandering water spirits who "roam the earth trying to make us listen, though . . . it's rather foggy as to what precisely we are supposed to hear," to cite Manohla Dargis's pan of the film in the *New York Times*. Nor did Dargis

FIG. 3.27. M. Night Shyamalan, *Lady in the Water*, ca. 47:06. The aquatic atmosphere imbues the narf's collection of everyday land objects with mystery.

appreciate the cinematography of the film by Christopher Doyle, "best known for his superb work for Wong Kar-Wai . . . [which] appears to have been lighted with a book of matches and a dying flashlight."[112]

Such fog, frustrating as it might be, is the signature of the underwater eye. In the film, it is justified not just by the pool but also by water in the air, including atmospheric mist and rain, water sprinklers that irrigate the lawn, and the fantastical water the narf brings from the pool. Further contributing to the wet feel of the atmosphere are other non-aquatic techniques of blurring, such as close-ups and dissolves. In contrast to suburban anomie, the wet optics of *Lady in the Water* throw a veil of mystery over everyday life. This mystery is in the spirit of the genesis of the film, which was a bedtime tale Shyamalan recounted to his daughters. The marine layer, however, is not enough to compensate for the weaknesses of the plot. Where the spell fails, the action appears sheerly comical, as in one scene shot through a window, in which the tenants glide through the aquatic mist armed with everyday objects, like brooms, mops, and sticks, in search of the beast who threatens the narf.

More persuasive is the installation revealed when the building's superintendent, the lovable misfit Cleveland Heep, played by Paul Giamatti, plunges into the depths of the pool and discovers the narf's lair. There, he finds a collection of random terrestrial objects—from rocks to antlers to glasses to old jewelry and butter knives (fig. 3.27). Backlit with Heep's flashlight, which enhances the haze, the collection takes on a neo-Victorian allure. Admiring the transfiguration that water works on such random objects, spectators are shown that all they may need is the appropriate emotional atmosphere to enchant the terrestrial everyday.

3.8. Tragic Patriotism, the Viewpoint of Drowning Soldiers and Sailors: *Saving Private Ryan* and *Pearl Harbor*

The underwater haze becomes a fatal shroud in war films with scenes of violent death at sea. Steven Spielberg's *Saving Private Ryan* (1998) opens in the cold dawn, amidst ocean mist, as it conveys the soldier-view of the D-day landing on Omaha Beach in Normandy. Along with subjecting spectators to the relentless, deadly German fire, as American soldiers beached their LCVPs, Spielberg explained, he wanted to convey there was no place to hide: "[S]oldiers weren't safe if they jumped into the water . . . four feet of water over your head could not stop a high-velocity shell from going right through you." To convey this vulnerability, cinematographer Janusz Kaminski took the camera underwater. He did so in "makeshift" conditions, as Spielberg noted. In order to make the water sufficiently transparent to show action, the filmmakers "basically dug a large hole, waterproofed it with Visqueen [plastic sheeting], and filled it up with purified water just to get these shots. . . . We had an underwater casing around the camera and a crane [set on a forty-foot flatbed trailer], so we were able to bring the camera up, submerge it, then bring it to the surface again. Instead of bullets, we shot pellets through the water that burst into blood bags."[113]

Spielberg's juncture of the camera above and below the waterline to sow confusion amidst violence imbues with tragic gravitas a zone that was a source of entertaining horror in *Jaws*. Further reminiscent of the carnage in *Jaws* are disjunctive views and clouds of blood. Indeed, Spielberg even echoes the blue, white, and red colors of shark carnage, although the palette becomes murkier in the cloudy Normandy water, filled as well with the green and brown uniforms of the dying soldiers. The red becomes their blood, while the blue and white become the colors of the sea and oxygen bubbles, which both enshroud them as they fall from the air, and that also result from the rocketing bullets.

Michael Bay amplified the hell of war at the waterline in his sequences portraying the Japanese attack on US battleships in *Pearl Harbor* (2001). While the movie in its entirety was not enthusiastically received, critics singled out the power of the bombing sequence. In the words of *New York Times* film critic A. O. Scott, the film conveys "disorder and mayhem on a large scale while maintaining a coherent sense of space and geography."[114] This coherence extends to the topside/underwater connection filmed by underwater photographer Pete Romano, a former navy diver, who had designed the SeaPar 1200-watt lighting used in Cameron's *The Abyss* together with Richard Mula. Romano also is known for Hydroflex, his company that designs innovative equipment for aquatic filming. As Romano explained to Pauline Rogers, he came to *Pearl Harbor* in the "middle of constructing HydroHead, his first pan and tilt underwater head." Romano recounts how HydroHead allowed the camera to "dive and breach with a full range of booms, pans and tilts, in, out and over the water," thus enabling the camera to access a varied vocabulary of views in the zone transitioning from the air to the depths. Working on location, as well as in a studio setting, Romano floated in "the uniform of the period, so that . . . I wouldn't stick out," while "several hundred extras took hits, were set on fire, and dodged bullets as planes flew overhead. It was almost a little too real."[115]

Amidst such dramatic conditions, Romano filmed extended sequences that take spectators into the struggles of drowning sailors, both trapped in the battleships and in the water. Furthermore, Bay created two iconic emblems of death that drew on underwater effects. The first is the inverted American flag shown from the bottom looking up (fig. 3.28). What was for Spielberg the shark's-eye point of view becomes in *Pearl Harbor* a tragic viewpoint from beyond the grave. The citation of *Jaws* extends to the soldiers' dangling legs, suggesting in this case the vulnerability of the human body to violence, no matter the strength of the individual. Sinking into the depths, the inverted American flag, riddled with bullets, turns spectral as well, its fabric translucent, and its colors bleached out.

The second emblem is the ship's name painted on the side of the USS *Oklahoma*, which Bay described as "the symbol of American might," which in the Pearl Harbor attack "just flipp[ed] over in six minutes."[116] Romano gave a terrifying immediacy to this event by showing it from the "point-of-view of a drowning sailor." To create this point of view, he utilized effects facilitated by the submarine ease of moving in three dimensions. "Half a dozen ex-Navy Seals dropped down below camera. As we rolled camera, they started floating up into frame. This made them look like they were dropping down into my frame, when they were actually rising to the surface. As they hit the middle of my frame, I and the camera spun upright keeping the legend in frame." Thus,

FIG. 3.28. Michael Bay, *Pearl Harbor*, ca. 1:42:25. The viewpoint of the drowning sailor.

this underwater sequence conveyed perceptual havoc to moviegoers, "as we rose up through dead bodies and panicked sailors trying to climb up on the hull of the Oklahoma."[117]

3.9. The Wave Spinning Air, Depths, and Surface:
Big Wednesday and *Point Break*

A sequence of shots at, below, and above the waterline signifies danger, but within the framework of adventure, in films about surfing. Such sequences make use of the shape of the wave that joins air, surface, and water when the wave hits the shallows and breaks. The ability to capture this continuity on film dates to 1968–69, when George Greenough, a "straw-haired autodidact kneeboarder from Santa Barbara," to cite Matt Warshaw, found a way to shoot from inside the enclosed space when the surface of the wave folds over and envelopes the rider. The surfer's experience in the tube of the wave produces an intoxicating sensation. To bring viewers into the space of this evanescent high, Greenough "built a nylon harness with a metal arm that" could hold a 20-pound camera, which he used in his independent film *The Innermost Limits of Pure Fun* (1970). Greenough positioned the camera just above shoulder level in the concluding section of the film, and he chose a 3.5 mm fisheye lens, which enabled him to show clearly subject matter that was extremely close, filming in "super slow-motion." Greenough often could not stop the film if the force of the wave would close over him and tumble him, as often happens riding in the wave's tube. As a result, "Greenough was the first person to capture the wipeout from underwater," producing, in the estimation of Warshaw, who had been a professional surfer before he became a surf historian, "a weirder, more vertiginous experience from a theater seat than in real life."[118]

The weird, vertiginous experience from the theater seat is created by the unique way Greenough joins the water's surface and the undersea. Greenough's sequences included in a single take the shiny form of the wave seen through air, the tense lines of force on the wave's surface, and a stunning blue-green atmosphere, which displaces the air on screen, as the camera rotates radically. The sequence also incorporates views of the wave as an insubstantial yet massive field of energy folding over, seen from below the surface, where its white breaking crest lightens the blue atmosphere as it impacts the surface. Greenough's captivating views fit the category of the sublime—dangerous for most viewers were they to be caught up in such intense energy yet conveying a sense of majesty from the comfort of the theater. Australian filmmaker David Elfick included both Greenough and his footage in *Crystal Voyager* (1973)— "the only surf film to show at the Cannes Film Festival." Warshaw quotes the ecstatic praise of a Melbourne film critic for this sequence, comparing "Greenough's work to that of Shelley, Keats, and Rimbaud."

The concluding sequence of *Crystal Voyager*, "Echoes," shot by Greenough, is set to an eponymous song by Pink Floyd written for the film, with appropriately submarine lyrics. "Echoes" opens with extensive traveling submarine shots of a wave and goes on to include in super-slow motion underwater shots of its froth, along with views of it breaking just above and below the waterline. The seamless continuity between topside and underwater in the energy of the breaking wave joins water and air within the circle of Greenough's fish-eye lens, suggesting the two atmospheres as making up our world. One year before the film appeared, NASA published the iconic blue marble photo taken in 1972 by Apollo 17, whose circle shows the earth as a water planet.

While *Crystal Voyager* attracted chiefly arthouse and surf audiences when it was first screened, both the surfer's view in the tube of the wave, as well as the wipeout sequence, were soon picked up by Hollywood. Director John Milius, who wrote the screenplay for *Apocalypse Now* among other credits, included Greenough's footage in the climatic episode of his coming-of-age surf film, *Big Wednesday* (1978), about three Southern California teenage buddies, Matt, played by Jan-Michael Vincent; Leroy, played by Gary Busey; and Jack, played by William Katt, who then go their different ways, set against the back-drop of the Vietnam War. "On the advice of Milius's pal Steven Spielberg," Patrick Pemberton explained, "the 'Big Wednesday' filmmakers sought to add big wave surfing scenes at the end of the film."[119] Spielberg has turned out to feature more oprominently in this book as a creator of underwater aesthetics than I anticipated when I started my research, due to his sensitivity to the ter-ror of views on, just above, and just below the waterline. Big waves, Spielberg realized, offered yet another canvas for his abiding interest.

Big Wednesday's climax shows the three buddies who reunite to surf "big Wednesday," the biggest swell to hit the California coast in living memory. The scenes of this swell were in fact filmed on the north shore of Hawaii, says Milius, "where we knew we could get really, really big waves," in dangerous conditions that required the skills of the world-class surfers who served as doubles, notably Ian Cairns, Peter Townend, and Bill Hamilton.[120] In such intense conditions, differences among the friends fade, and they renew their camaraderie, first to master the waves, and then to save Matt, when he achieves a ride on huge wave, wipes out, and almost drowns. The epic wipeout se-quence montages surfing shot from the beach; signature footage by Gree-nough, who is credited in the "Special Water Unit," along with Dan Merkel; and additional turbulent, handheld underwater footage of churning water and bubbles, including bodies in the depths, tumbling out of control. The novel perspective of the wipeout from underwater belongs to the aesthetic of the sublime. Notably, the turbulence and impact of the lip of a big wave crashing into the depths as seen from below gives a heightened sense of its violence and might.

Expectations for the success of *Big Wednesday* were high, and Pemberton writes, "Spielberg and George Lucas famously traded 'Big Wednesday' profit points with their next movies, 'Star Wars' and 'Close Encounters of the Third Kind'—a move that would bring Milius millions."[121] However, Milius did not create a blockbuster look for the film. A flop when it was released, *Big Wednes-day* would win high praise within surf culture over the next decades. Further, the fascination of the breaking wave seen from the depths would be picked up subsequently, in surf films and also in films for general audiences. In Kathryn Bigelow's box-office hit *Point Break* (1991), for example, the wipeout footage aligns with the point of view of Johnny Utah, played by Keanu Reeves, when he almost drowns while trying to teach himself to surf.

The use of the wave to join air, waterline, and depths would travel beyond surf films to become so iconic that it featured in the signature montage

FIG. 3.29. Andrew Byatt and Alastair Fothergill, *The Blue Planet*, ca. 0:30. Signature image: a great wave as if seen through water, in a frame whose shape suggests the blue marble and a fish-eye lens.

opening each episode of the first BBC *Blue Planet* series (2001, US airing 2002). This brief sequence ends with a few seconds of powerful waves seen both underwater and topside. Then, a giant wave takes shape in the circle of a fish-eye lens, whose form also recalls the blue marble photo. While the wave rises majestically in the air, it is steeped in the murky blue atmosphere of the undersea (fig. 3.29).

3.10. The Wreckage of Civilization Submerged:
Waterworld and *Titanic*

The vision of a majestic civilization engulfed in a flood dates in the West to antiquity, most famously with the story of the lost Atlantis, first recounted by Plato in the fourth century BC. In Verne's playbook of submarine fantasy, he imagined that undersea exploration would further nautical archaeology, and thus, *Twenty Thousand Leagues under the Seas* includes a memorable scene where Nemo takes Aronnax to view the ruins of Atlantis, which Nemo indeed has located. The underwater film set enabled creators to show architectures of sunken wreckage, amplified beyond the ship to other types of constructions. Jacques Tourneur's *War-Gods of the Deep* (*City in the Sea*) (1965), inspired by Edgar Allan Poe's poem "The Doomed City" (1831), fantasized a submerged city, although in Tourneur's film, its inhabitants are still alive, thanks to a science fiction supply of air, until the volcanic eruption that destroys the sunken city at the film's end.[122]

In contrast, Kevin Reynolds took spectators down for a post-apocalyptic visit to ruins in *Waterworld* (1995), an aquatic variation on the *Mad Max* films, including their junkyard aesthetic. In *Waterworld*, the melt from polar icecaps has submerged the earth. The survivors live on islands they create from wreckage, and travel across the waters, desperately bartering and warring for the scant resources remaining that support human life. The film's protagonist is a survivor evolved to inhabit the two realms of air and water. Kevin Costner plays the mutant creature with gills behind his ears named the Mariner, revising the monstrous qualities of *Creature from the Black Lagoon* into a romantic lead. The Mariner belongs to a revisionist lineage celebrating Gill-man, including most recently Guillermo del Toro's *The Shape of Water* (2017), which transforms the Amazonian monster into an empathetic, misunderstood creature, tortured by racist white scientists who subjugate and despise the Global South.

In *Waterworld*, underwater sequences furnish lyrical visuals that unlock emotion, in brief respites from the instrumental brutality of human relations on the surface. On a lighthearted note, the Mariner rescues his romantic interest, Helen, played by Jeanne Tripplehorn, from marauders, led by Deacon, played by Dennis Hopper, by hiding her underwater. He passes oxygen that he can extract from water to her by kissing her, in a rom-com variation on buddy breathing, a staple of underwater escapes since *Sea Hunt* (fig. 3.30). Embracing in the rich aqua-blue of the liquid sky, the pair are viewed by the camera from below, silhouetted, as if they were flying.

This impromptu dive is a moment of respite, during a topside battle scene when Deacon and his associates kidnap the child adopted by Helen named Enola, played by Tina Majorino, and destroy the Mariner's boat. This violence occurs just after a melancholy descent when the Mariner takes Helen in a steampunk diving bell to see the fabled "Dryland" that survivors imagine must exist somewhere amidst their aquatic dystopia. The Mariner has found Dryland, in contrast to its name, on the deep seafloor, which is where the Mariner, enabled by his genetic mutation, collects the earth he barters for survival above the surface. In another ironic twist, the Mariner guides the diving bell into the depths by the light of flares, made from dynamite sticks, that he holds outstretched like a torchbearer. In contrast to the opening of *The Silent World*, this plunge does not lead to a new frontier or new technologies, but rather reveals ruins. The bell is tethered to what is presumably a battered depth gauge

FIG. 3.30. Kevin Reynolds, *Waterworld*, ca. 1:29:32. The Mariner and Helen buddy breathing.

at the surface, which shows the depth as around 750—feet or meters, we don't know which.

There, together and yet separated, the Mariner breathing water and Helen looking through the glass window of the bell that lets her breathe air, the two contemplate disaster film images of a devastated metropolis. The fantasy includes a submarine crashed into a city, which layers warfare upon environmental apocalypse. The views of these ruins are steeped in the gothic mood of Cousteau's shipwrecks, including haze, obscurity, and a deep blue color palette, illuminated by the orange-white light of the flare (fig. 3.31). Such wreckage—on the scale of a civilization—does not inaugurate an adventure within this action film. Rather, it suspends action, freezing the spectators' attention on these images of catastrophe sealed off from history, submerged in time and space. This traumatic fixation recalls the emotional paralysis when reliving trauma described by Maurice Blanchot in *The Writing of the Disaster*. Blanchot muses, "*Always returning upon the paths of time, we are neither ahead nor behind, late is early, near far,*" since disaster renders time jammed and inoperable.[123] Disaster freezes time in a stasis Blanchot represents as time's spatialization. In Verne's *Twenty Thousand Leagues*, Atlantis partakes of such stasis and spatialization, and Dryland, too, is a location that the Mariner can revisit and view at his leisure.[124] And yet, in a survivalist twist on Blanchot's paralysis, the Mariner also has been putting the site to use to eke out his existence, extracting from it earth that is highly prized in the flooded world. This potential does not

FIG. 3.31. Kevin Reynolds, *Waterworld*, ca. 1:24:38. The sunken Atlantis of modernity, aka "Dryland."

negate the representation of catastrophe, and yet still, it affirms, albeit at the level of bare subsistence, the human effort to endure.

Waterworld, however, did not etch a memorable image of sunken modernity on spectators' imaginations. In *Titanic* (1997), in contrast, James Cameron captivated the attention of the world with the story of the most famous shipwreck of the twentieth century. The RMS *Titanic* sank on its maiden voyage because of its engineering flaws and the arrogant navigation of its hard-driving captain. In this disaster, around 1,500 people drowned, disproportionately working class, because the shipping company failed to provide lifeboats for all passengers and crew members on board. Critics have justly taken *Titanic* to task for its nostalgic transformation of capitalist exploitation into a Romeo and Juliet–like blockbuster, emphasizing the sacrifice of the working-class hero, Jack, to save his upper-class love, Rose.[125] Alexandra Keller characterizes the public fascination with this film by using a line from Guy Debord on the spectacle: "It is the sun which never sets over the empire of modern passivity. It covers the entire surface of the world and bathes endlessly in its own glory." Keller continues, "Debord was defining the spectacle in those lines, but he might just as well have been speaking of *Titanic*, its maker, and its format, the blockbuster."[126]

What such a reading of the film misses is the remarkable first twenty minutes of the film, including a sequence taking viewers to the wreckage, which was first located off Newfoundland by oceanographer Robert Ballard in 1985.[127] Keller describes this sequence as "among the densest representations of representation and spectatorship on film."[128] Among the concerns it works through is the dense entanglement of both documentary and narrative film on the underwater film set. Both documentary techniques and narrative imagination, Cameron suggests in this sequence, are needed to reveal the extraordinary and also terrifying realities hidden in the depths.

The episode opens with two Mir submersibles, which are the supports for a powerful lighting system permitting filming of the wreck amidst obscurity. This lighting system was first essayed for the IMAX documentary *Titanica* (1992), directed by Stephen Low. To provide enough illumination, the two Mirs used the HMI SeaPar lights designed for Cameron's *The Abyss*, initially meant to function down to 250 feet, which were adapted so that the lights were able to work during dives "to the bow and stern sections [of the wreckage] at a depth of 4,000 meters."[129] After treasure hunter Brock Lovett, played by Bill Paxton, creates a cheesy narrative filming through a porthole of the submersible, the treasure hunters deploy Snoop Dog. Snoop Dog is an remotely operated vehicle that Cameron had designed by Western Space and Marine, the company that made the futuristic helmets for *The Abyss*. Cameron had imagined such ROVs in that film, where they were named Big Geek and Little Geek, and Snoop Dog was initially conceived as a prop.

However, Snoop Dog was more than a prop and captured footage from inside the wreck site used in the frame episode introducing the film's reconstruction of the disaster. This footage includes the previously mentioned "silt streaming through intricate bronze-grill doors," and "a woodwork fireplace with a crab crawling over the hearth," as well as "luxurious suites overgrown with deep-sea animals."[130] In Cameron's words, the exhibition "set a level of excellence for the rest of the movie. . . . Sets and costumes had to live up to the example set by visiting the real wreck."[131] At the same time, despite Cameron's insistence that documentary preceded fantasy, in fact documentary and fantasy were mixed in the sequence that reveals the haunting secrets of dwelling quarters, once cabins, now graves. Studio images include a single worn woman's boot prostrate on the floor, a fragmented pair of glasses, and a porcelain mask in the sand with the proportions of a child and empty holes for eyes. These images were integrated with actual views; thus, the woodwork of a studio fireplace in a cabin is montaged with a view of it captured from inside the wreck (figs. 3.32, 3.33). With the imagined treasure hunters' visit to the real *Titanic* as prelude to a fictional plot, Cameron is creating his version of

FIG. 3.32. James Cameron, *Titanic*, ca. 6:41. Ornamented studio mantle on the wreck set of *Titanic*.

FIG. 3.33. James Cameron, *Titanic*, ca. 6:44. Underwater capture of cabin ornament in the wreck of the RMS *Titanic*.

docudrama, to use Malle's term from *The Silent World*; however, it is a type of docudrama where real technology and imagery intensify the significance of Cameron's historical fiction, rather than a docudrama that uses aesthetics to intensify documentary. In yet another turn, with the invention of actual deep-sea ROVs, modeled on fantasy ROVs from *The Abyss*, narrative film anticipates and, moreover, helps to shape technical innovation essential to filmmaking in remote depths.

For all its dense reflexivity, this opening sequence on the wreck continues the uneasy mixture of melancholy and strange anticipation we have already seen with the portrayal of the gothic wreck in the films of Cousteau. Cameron shows

FIG. 3.34. Jacques-Yves Cousteau, *Sunken Ships*, ca. 19:53. Absurdity: documentary image of a diver sitting in a sunken bathtub.

the *Titanic* as an awe-inspiring ghostly site, even if its sublimity is mocked by the treasure hunters in the story. The image of a bathtub in a sunken ship had fascinated Cousteau from the first film made with scuba, *Sunken Ships*. There, the narrator interprets this relic playfully, as the mark of "people's small, daily preoccupations," and a diver fits himself into it (ca. 19:53) (fig. 3.34). When Cameron shows the bathtub of the RMS *Titanic*, a lone fish swims by. Sighting the bathtub in *Titanic*, the geeky engineer remarks, "oops, somebody left the water running" (ca. 7:29). With such uneasy sarcasm, the engineer recalls the might of the ocean that no human vessel can contain, and the lives swept away (fig. 3.35).

At the same time, the wreckage lost for decades and then reawakened has an unsettling afterlife that is hard to condense into a single word. When Cameron described visiting the wreck site with his ROV "avatar," he compared it to "out-of-body experiences of ghostwalking." Surrealism was another aesthetic that came to Cameron's mind when he sought to express his reaction to this "gothic ruin [that] exists now in a ghostly limbo, neither in our world nor completely gone from it. The rusticles have transformed Edwardian elegance into a phantasmagorical cavern, a surreal underworld."[132]

FIG. 3.35. James Cameron, *Titanic*, ca. 7:28. Tragedy: sunken bathtub on the wreck set of *Titanic*.

3.11. Sea of Fantasy: *The Life Aquatic with Steve Zissou*

Wes Anderson's *The Life Aquatic with Steve Zissou* (2004) celebrates the underwater environment as the generator of delightful aesthetic effects, at a far remove from Cameron's emphasis on documentary views as setting the standard for simulation in his comments about filming the RMS *Titanic*. Rather than incorporating documentary into narrative cinema, *The Life Aquatic* exposes documentaries as fanciful constructions, with a loving jab at undersea documentary's most famous icon, Cousteau. The plot of Anderson's film turns around a quest for a mythical killer shark, the jaguar shark, which references "Sharks," directed by Philippe Cousteau Sr. and Jack Kaufman, the first episode in the TV series *The Undersea World of Jacques Cousteau*, which premiered on January 8, 1968 (after a 1966 pilot) and would go on for eight years of memorable viewing. Anderson cites with zest, piling on details from the TV series in general, and the shark episode in particular, down to the aerial shot showing the Institut Océanographique in Monaco that Cousteau directed (1957–88), which Anderson echoes in establishing Steve Zissou's base at the opening to the film. Zissou too has a yellow submersible recalling the Anorep 1, the yellow deep-sea submersible first used by Cousteau in 1966, the same year the Beatles released their single "Yellow Submarine," and suggestively similar to the playful cartoon submersible in the Beatles' film *Yellow Submarine* (1968).

The figure of the shark is so imbued with legend that it offers perfect conditions for Anderson's transformation of the underwater environment into a sea

of fantasy. The film roams across shark filmography, not forgetting *Blue Water,
White Death* and *Jaws*, along with the oeuvre of Cousteau. References to all are
appropriately made in a film-within-a-film that Zissou, played by Bill Murray,
screens at the opening to *The Life Aquatic*, showing his first encounter with a
predator that kills his friend Esteban. While this fictional film cites conventions of shark carnage, they have gone awry. Thus, it starts with a cloud of
blood and passes on to out-of-focus shots, including an awkward pan from
Zissou to Klaus, played by Willem Dafoe. These amateurish views remain resolutely topside, in an exaggeration of the *Jaws* technique of withholding knowledge of the killer to provoke suspense. There is no comparable shark's-eye
point of view for the reverse shot of killer to suggest it even exists—and Zissou's quest for the storied shark begins.

In a further gesture of reflexivity, Anderson draws attention to directorial
craft—or lack thereof—on the underwater film set, as he does in other settings
throughout his oeuvre. At the press conference launching the plot, after the
screening of the film where the unseen shark kills Esteban, Zissou is asked a
question in Italian, which the translator phrases as, "Was it a deliberate choice
never to show the jaguar shark?" Zissou replies, "No, I dropped the camera," to
audience laughter (ca. 3:52). If *Jaws* took so long to show the shark, it was, as I
have mentioned, in part owing to a malfunctioning machine. In *The Abyss*, we
first see an NTI craft with Lindsey, who is piloting a submersible, but her failure
to capture an image on their first sighting creates suspense. "So you didn't get
anything on the cameras?" Bud asks. "No," says Lindsey, "I didn't get a picture
of it." "What about the video?" Bud continues. "No. We lost power right then"
(ca. 34:12). Through his intertextuality, Anderson reminds us of the importance of
pictorial documentation for imbuing undersea creatures with a sense of reality—
all the more so in the case of artificial, if not mythic, creations like Spielberg's
mechanical shark, Cameron's NTIs, and Anderson's own fantastic marine
kingdom.[133]

"Tell me something, does it actually exist?" smugly asks Captain Hennessey,
played by Jeff Goldblum, Zissou's hypersuccessful competitor and "nemesis,"
in Zissou's words (ca. 6:02, ca. 6:30). "Does the jaguar shark exist in the (film's
imaginary?) undersea?" is one interpretation of this line; another is, "What is
existence for the film spectator, enjoying the reality of the simulation?" Even
documentaries simulate events, although they abide by the ethos that the event
could plausibly happen. When Anderson's shark appears, spectators savor its
fantastical existence. The shark is an eight-foot puppet filmed with stop-motion

FIG. 3.36. Wes Anderson, *The Life Aquatic with Steve Zissou*, ca. 2:33. Steve Zissou with a school of fluorescent snappers at the premier of a film-within-a-film.

animation, created by Henry Selick.[134] When Selick explains, he gives another version of the fun Anderson is having with "real" undersea creatures in narrative film: "Wes realized that if he was going to make a movie about an underwater adventure, he was going to need a menagerie of underwater life. And he looked to me, because he didn't want to use computer-created or -enhanced digital effects. He wanted something *that looked more real*, as ridiculous as that sounds."[135] Selick says about the puppets for the stop-motion, "We only made a few, but we animated them many times at different angles. They were multiplied in compositing. We just animated them at so many different speeds and so forth. Rather than totally switch over to CG when nothing else was CG, it made sense to hand-animate a very small school and then cut and paste that."[136] Given its dimensions, according to Sean Hutchinson, "the stop-motion animation on the Jaguar Shark was done upside down to make it sag against gravity."[137]

These creatures were tendentiously old-school, pun on "school" intended, such as the "fluorescent snappers [that] unexpectedly appear in the shallows, extremely rare at this depth," part of the "documentary" premiered in the film's opening (ca. 2:28) (fig. 3.36). Selick recalls, "We definitely had a lot of fun doing the film." Some of the animated creatures took their inspiration from real creatures, such as the dolphin "albino scouts," but they were embellished, an ornamental human touch often signified with zany color. Thus, the film contains creatures such as the crayon ponyfish, sugar crabs, and a paisley-colored octopus. In addition, Selick recalls, "we came up with a lot of things

FIG. 3.37. Wes Anderson, *The Life Aquatic with Steve Zissou*, ca. 1:47:48. The jaguar shark's jaws from the spectators' point of view.

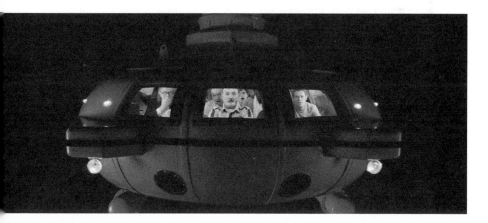

FIG. 3.38. Wes Anderson, *The Life Aquatic with Steve Zissou*, ca. 1:47:31. The spectators from the jaguar shark's point of view.

that wound up on the cutting-room floor, including the Rat Tail Envelope Fish, this creature that can turn itself inside out."[138]

In Selick's animation, the jaguar shark is captivatingly bioluminescent and obviously artificial (fig. 3.37). Anderson further reminds us that the undersea world comes to us mediated through human projection in the scene of the shark's sighting. The members of the expedition do not encounter the jaguar shark diving but rather as it passes before them when they are arrayed in rows in the deep-sea vessel, ~~Jacqueline~~ *Deep Search*,[139] as if they are spectators at the

FIG. 3.39. Jacques-Yves Cousteau, *The Undersea World of Jacques Cousteau*, "Sharks," ca. 5:34. Cousteau with a model of a prehistoric shark at the Institut Océanographique, Monaco.

movies, with rapt gazes seen also from the jaguar's shark's fantasy point of view (fig. 3.38). The shot/countershot convention features hence foreground the pleasures of the spectacle, both for the humans, and, it would seem, the shark that examines them, although the scene does not create an exchange of looks between the jaguar shark and the vulnerable humans, perhaps because it is too majestic and mysterious. The shark, too, recalls movies when it glides by again, its massive body broken up by the vessel's windows, as if the shark were being exhibited across the frames of an analog film strip.[140] With his keen eye for the surreal amidst the serious, Anderson's animated jaguar shark perhaps also both parodies and improves upon the facsimile of a massive prehistoric shark hanging in the Institut Océanographique of Monaco at the beginning of the episode "Sharks" in *The Undersea World of Jacques Cousteau*. While Cousteau's Institute is pursuing cutting-edge deep-sea research around the world, its director at the opening to the episode "Sharks" does indeed pose seemingly in earnest before a huge, hokey model (fig. 3.39).

"We are phonies, so we can make it better," Browning told Tors, in making a pitch for why his Hollywood production team rather than Cousteau should

FIG. 3.40. "On the Underwater Filmset of *The Life Aquatic*." Photograph courtesy of Hydroflex, Inc. Underwater director of photography, Pete Romano, filming amidst the set's exuberantly fake seaweed in the episode "Investigating the Phantom Signal."

shoot the underwater sequences for Terence Young's *Thunderball*. *The Life Aquatic* is making a similar point, with the added twist that its artifice is a form of humanism. Even beneath the sea, in Anderson's film, both filmmakers and their characters cannot help but affirm their flawed vitality, whether it is through the imagery they construct or inconsistent and sometimes incoherent revelations. Tellingly, the human actors are about the only biological creatures in the film's underwater scenes. These sequences are filmed entirely in a pool set, where even the seaweed is textile and the marine creatures are animated and fantastically, if humorously absurd (fig. 3.40). There is one more physical element that Anderson does not fabricate amidst the film's pervasive artifice: its submarine scenes are shot wet for wet. Even Anderson acknowledges the unique, if resistant, properties of water, at once physical nature and too poetic to embellish with his signature artistry.

Epilogue

My study concluded with *The Life Aquatic with Steve Zissou* to underline that even for a filmmaker devoted to artifice such as Anderson, the visuals he can achieve in water are too alluring to merit improvement. The aquatic atmosphere helps to form all fantasies considered in *The Underwater Eye*, from documentaries to the most whimsical films. This point is so obvious that it hardly bears mention: the physical features of our planet are material factors that play a role in expression, at least in all live-action films. While for everyday situations, spectators take these physical features for granted, their presence stands out in depictions of extreme, toxic, or rare environments, as does the formative role played by the possibilities and limitations of technologies for their capture.

In considering the formative role of the underwater environment on moving imagery, I have attended solely to aesthetics. Even when considering documentary films, I have focused on crafting, artistic traditions evoked, and the emotions they solicit. I have not addressed the relation of such imagery to the real locations they depict, nor asked what they choose to show of such locations—and what they left off screen. I did not raise this question imperative for the study of documentary as a genre because to answer it would have made my project unmanageable. Drawing limits is an important principle of scholarship, particularly when delving into understudied topics.

To ask how documentary imagery is true to the reality referenced would require broaching the crucial if complicated question: Just what reality does a documentary represent? As film critics have pointed out, there are multiple documentary ethics, ranging from films that record the trace of a phenomenon's physical presence to those that convey the quality of subjective perception to those disseminating a message.[1] The documentary corpus, moreover, is robust, notably starting in the 1950s, owing to the same dive and film technologies that enabled narrative film. The post–World War II fascination with the oceans as a new planetary frontier stimulated the production of undersea documentaries for general audiences, as did the variety of screening venues that emerged, from oceanariums and natural history museums to television. Yet another body of work it includes are crossover scientific films that were distributed beyond the laboratory.[2] The documentary corpus flourishes yet more expansively in the twenty-first century. Digital photography gives filmmakers exponentially increased recording hours to capture elusive wildlife. Virtual platforms, notably social media, provide additional venues, and many more people have access to excellent, inexpensive cameras that enable them to achieve footage that would have been costly and difficult in the analog era.

In this epilogue, I want nonetheless to give an example of how documentary creators also craft their imagery by making use of aesthetics distinctive to the aquatic environment. My example will be a documentary film in the message-making subgenre: the natural history seven-part television series *Blue Planet II* (2017), the most popular moving image spectacle about the oceans in the past decade, seen by over a billion people around the globe.[3] As I explore its dramatic imagery, it will emerge that even as *Blue Planet II* filmmakers display novel scenes of submarine reality, they organize their imagery according to the underwater aesthetics studied throughout this book. This section thus returns to the complications that submarine exhibition poses to the distinction between documentary and narrative film, addressing it now from the perspective of how documentaries reference reality.

A sequel to *The Blue Planet* (2001), the BBC's *Blue Planet II* premiered in 2017 on a Sunday night at 8 p.m., sixty-one years after the BBC's *Diving to Adventure*, which launched television miniseries exploring the depths. Both *The Blue Planet* and *Blue Planet II* fit the documentary category that Bill Nichols defines as the expository mode—which "directly addresses issues in the historical world."[4] *Blue Planet II* challenges its viewers to counteract the destruction of the oceans' health by human industry, with an emphasis on plastic pollution and fossil fuel emissions. This challenge was heeded, notably, in the United Kingdom, and indeed around the globe (regrettably, less in the United States), prompting consumers to turn away from single-use plastic. Indeed, *Blue Planet II* "is considered one of the most influential series of its kind," according to Helen Wilson.[5] It has been "cited repeatedly in policy announcements and calls for action, including a high-profile speech by the [UK] Prime Minister in January 2018."[6] I hence choose

Blue Planet II not only for its popularity but for its impact and contemporary relevance. Raising public awareness about climate change is a top priority for citizens and policymakers today—and environmental documentaries like *Blue Planet II* suggest that creative expression can play a useful role in this task.

In advertising *Blue Planet II*, the BBC promoted the series for "some of the most spectacular events and compelling animal characters in nature," thanks to "the latest diving and submarine technologies," making it "possible to explore the oceans today like never before."[7] From the vantage point of *The Underwater Eye*, what stands out is how the show's creators shaped such never-before-recorded and in some cases never-before-seen ocean life along the lines of preexisting submarine fantasy, including the liquid fantasies discussed in part 3. Audiences too were captivated by echoes of liquid fantasy, from episode 1, "One Ocean," which premiered on October 29, 2017.

"[T]he scene everyone's talking about," according to an online article in British *GQ* dated the same day, was the segment "when giant trevally fish jump out of the sea and eat seabirds mid-air."[8] The giant trevally is a fish that on its own is worth documentation, from the perspective of Dan Beecham, the *Blue Planet II* cameraman who captured underwater views of these powerful hunters for the episode. "They have an incredible presence and perform many extraordinary behaviours, including gathering in huge groups to swim up estuaries on South Africa's east coast and they have even been documented in scientific literature attacking and killing reef shark."[9] Among the "extraordinary" behaviors, the trevally segment focuses on is a feat that had only been rumored: the fishes' ability to take down a bird in flight. As producer Miles Barton recalled, the film crew flew to the remote Farquhar atoll in the Seychelles in the Indian Ocean, "with hundreds of kilograms of highly expensive kit with no real confirmation." "But the punt was worth it, and the incredible footage is the first of its kind."[10] This incredible footage captured giant trevally hunting birds, shown from the depths, the water's surface, and the air.

To organize these diverse perspectives, editors used the liquid fantasy of shark carnage. Spectators were the first to note the echo of *Jaws*. "Don't go back in the water," the popular website NME.com subtitled its review of the episode that appeared the following day.[11] The word "horror" recurred in reviews after the episode aired, couched in gleeful purple prose. The *Daily Mail*, in previewing the episode, called the giant trevally a "monster [that] tracked the flight-path of the tern before intercepting and eating it . . . in one gulp."[12] "If you need an idea for a Halloween costume, why not dress up as a giant trevally?" asked the *Telegraph*.[13]

FIG. 4.1. James Honeyborne and Mike Brownlow, producers, *Blue Planet II*, episode 1, "One Ocean," ca. 13:38. The giant trevally, a "monster [that] tracked the flightpath of the tern before intercepting and eating it . . . in one gulp" (Lambert, "Stunning Moment").

When *Blue Planet II* transferred shark carnage to wildlife documentary, Cape Cod summer crowds became flocks of sooty terns, a seabird that touches down on land to breed. Like *Jaws*, the segment starts by showing a convivial, vibrant community. Vacationers are replaced with raucous parents and chicks, gathered on white sands at the edge of an exquisite aqua lagoon. The segment then focuses on that tense yet exciting moment for human parents: when a fledgling launches out on its own, in this case taking flight over the lagoon's protected waters. Spectators send vicarious wishes for success, as a chick struggles, gets itself airborne, and tires, bobbing on the water to rest. Suddenly, with a shocking, unexplained swoosh, the chick is pulled under and disappears. This violence without a visible cause recurs three times, until in the next attack, we see a large fish head break the surface, and the narrator declares: "giant trevallies" (ca. 13:42) (fig. 4.1). Like the inlet in *Jaws*, even a lagoon is not safe with such fearsome creatures on the prowl.

In the next few minutes, the film calls on familiar shots from *Jaws*, within the much more compressed timeframe of just one segment in an hour-long episode. These shots include underwater views of the predator tracking its prey and topside views of calm with no indication of the giant trevally's presence. We see a surface view of the trevally's dorsal fin cutting through the water, and handheld camera movements at the waterline when the trevally attacks. The episode includes a number of shots just below the waterline of the trevallys, and in one, a trevally brings its eyes out of the water. In place of legs dangling underwater seen from a shark's-eye perspective, we share a fish's-eye point of view of the air, looking up at the flying chick enwatered through the lens of the sea. This shot occurs as the sequence builds toward its dramatic counterpart to the culminating indictment of shark monstrosity in *Jaws*: the ability of a predator to transgress the bounds of its natural element. *Jaws* captured the shark attacking in air by using the mechanical shark Bruce; the topside camera person in *Blue Planet II* used the hundred thousand–dollar Phantom Flex, a 4K digital camera capable of filming one thousand frames a second. Shown in super slow motion, the fish takes down a few chicks from the air, until one strike when a chick breaks free. As the chick flaps away, the balance of the elements is restored, and the segment concludes with the apex predator falling back into the water, spinning, with the camera catching its glassy eye.

Following trevally carnage, the first episode takes us to schools of mobula rays swimming at night lit by dazzling bioluminescence, caused by "[n]octiluca scintillans, or sea sparkles, that light up with the rays' wing beats."[14] As executive producer James Honeyborne explained, the film's ability to capture this imagery in remarkably high definition depended on digital cameras able to film "in the dark, in low light, in color at 4K."[15] At the same time, he remarked on the aesthetic power of the scene. "It was like a Disney film," Honeyborne said, perhaps thinking of the bioluminescent nighttime fairies dancing in *Fantasia* (1940).[16] In contrast to the horror shaping of the trevallys, filmmakers organized this scene of predation as an underwater dance, a trope that has been used to celebrate aquatic movement across the history of underwater film. The dance in the mobula ray sequence is an "extraordinary ballet of life and death" to use the words of Attenborough's narration (ca. 21:46) (fig. 4.2). Audiences too were captivated by the dance when the episode first aired. Indeed, the popular website Mashable went so far as to include the ray sequence among the ten best moments in the entire series, in an article by Isobel Hamilton describing "[m]obula rays swimming through bioluminescent plankton [that] cause it to glow, resulting in footage that looks like a scene from *Avatar*."[17] I did not find an invitation to take pleasure in the killing in fact depicted in this lyrical scene. Nonetheless, the rays' silhouettes, ornamented with glitter, recall, to this viewer at least, the dance of hunting

FIG. 4.2. James Honeyborne and Mike Brownlow, producers, *Blue Planet II*, episode 1, "One Ocean," ca. 22:17. The mobula ray's "extraordinary ballet of life and death" (ca. 21:46).

FIG. 4.3. Terence Young, *Thunderball*, ca. 4:53. Maurice Binder's credits mimic a bioluminescent dance of hunter and hunted.

featuring gliding silhouettes amidst luminous trails of bubbles that Maurice Binder designed for the opening credits to *Thunderball* (fig. 4.3).

Hamilton's reference to Cameron's film is telling, because in nighttime scenes of *Avatar* (2009) on the planet Pandora, Cameron fosters a sense of wonder by using a palette contrasting "violet-black" darkness with biolumi-nescence, similar to the palette of wonder found in the abyssal depths.[18] Because *Blue Planet II* shows the mobula rays feeding at night, filmmakers find in shallow water the beauty of such deep-sea contrast, although the reference to Cameron in this scene is low key. *Blue Planet II* explicitly evokes *The Abyss* in the subsequent episode 2, "The Deep," which aired on November 5, 2017. Coming along as vicarious passengers on a deep-sea submersible, spectators travel to what Attenborough calls "an alien world" "in the Pacific 200 meters down" (ca. 5:28).[19] When the episode descends from "The Twilight Zone" to "The Midnight Zone," Attenborough tells us, "there's life . . . but not as we know it" (ca. 12:11). "Down here, in this blackness, creatures live beyond the normal rules of time" (ca. 14:46). Thus, the "siphonophores . . . virtually eternal . . . repeatedly clone themselves" (ca. 14:52). The "[a]lienlike creatures" (ca. 12:28) signal with a "language of light" that defies our understanding (ca. 12:50). Siphonophores may lack the intelligence of Cameron's NTIs, but their bioluminescent flashes in the velvety darkness are beautiful to behold.

The following episode, "Coral Reefs," first shown on November 12, 2017, continues *Blue Planet II*'s presentation of documentary reality shaped along the lines of fantasy. As the narrator explains, this episode organizes reefs as "underseas cities crammed full of life," zones of "opportunity," filled with "fierce rivalry for space, for food, and for a partner" (ca. 0:55). In the 1950s and '60s, coral reefs were marketed to tourists as places of otherworldly beauty, or as nature's living aquariums.[20] The imagination of the coral reef as an underwater city or at least a sprawling suburb dates to Pixar Studios' *Finding Nemo* (2003), directed by Andrew Stanton and Lee Unkrich. While I did not include this film as a liquid fantasy, since it achieved underwater imagery through animation, the creators of *Finding Nemo* consulted marine experts in developing the film. The result was a fantasy inspired by undersea conditions, if not made of them. Animators crafted imagery respecting human optics in the depths, including the behavior of light and views of moving fish bodies, even as they modeled fish on people, drew a toothsome shark resembling Bruce in *Jaws*, and brightened the undersea with a Pixar palette.[21]

In 2017, new camera and dive technologies enabled filmmakers to capture more extensively than previously the actual life of coral reefs, "from penthouse suites to backstreet dens" (ca. 5:27). Camera innovations include a miniature wide-angle lens small enough to penetrate into a reef's interstices. Dive innovations include rebreathers that did not release bubbles or make noise and thus enabled divers to lurk unobtrusively for up to four hours at a time in shallow water, helping creatures "just relax and let you into their world."[22] Along with the familiar theme of predation often used to organize wildlife films, "Coral Reefs" showcased symbiosis in its live-action realization of

Finding Nemo's undersea sociability. We see turtles seeking out a rock that is the habitat of blennies and surgeonfish, who groom them, cleaning off "algae, parasites, and dead skin" (ca. 14:36). Members of a single species cooperate as well, when a group of male clownfish use teamwork and their snouts to move a coconut shell beneath an anemone, creating a surface under the anemone's shelter where the female places her eggs. While monocle breams have few defenses, they work together to mark the underground burrows of a predatory meter-long marine worm with fearsome jaws, *Eunice aphroditois*, also known as a bobbit worm. As BBC Earth archly remarked, "[T]he creature was given its sinister nickname in a 1996 field guide, in reference to John and Lorena Bobbitt (millennials who may not have heard of them—look it up)."[23]

Blue Planet II's attention had been drawn to this behavior of the monocle bream in part by an article by scientists Jose Lachat and Daniel Haag-Wackernagel titled "Novel Mobbing Strategies of a Fish Population against a Sessile Annelid Predator."[24] When night falls on the reef, the "sessile annelid predator" turns into "the worst creature that ever existed" (*Independent*) and "the stuff of nightmares" (*RadioTimes*).[25] With the assistance of a new "one of a kind" infrared lighting system," the BBC filmed the worm's activities in the dark.[26] The unsuspecting prey is a lionfish, itself a local predator, which is now an invasive species throughout much of the world. Primed to follow the lionfish hunting, viewers are startled when it passes over a somewhat indistinct shape that seizes it, in a brusque jump scare, punctuated in the soundtrack with a chord. The lionfish flails in darkness and clouds of sand, with just its writhing tail sticking up, before it disappears into the seafloor. "The hunter has become the hunted," declares Attenborough (ca. 25:48). Along with the obscurity of the atmosphere used by *Jaws*, and the recesses of a wreck where a moray eel lurks in *The Deep*, the bobbit worm sequences add a trapdoor to the underwater set of horrors. As with other liquid fantasies evoked by the series, spectators when the show first aired were enthusiastic about the scene's horror house conventions. *RadioTimes* commented, "The 'giant carnivorous' bobbit worm turns Blue Planet II into an actual horror movie," including a thread of Twitter feed where "horrified viewers" "share[d] their dismay."[27]

Yet another liquid fantasy recurring in *Blue Planet II* is the kinship between marine creatures and humans. One visual expression of this fantasy that has presented challenges for submarine filmmakers is imagery that creates continuity between the atmospheres of water and air, given their different physics and hence the different demands on recording. As we have seen from *The Silent World's* dolphin sequence, artful montage of topside and underwater views solves this problem and creates a sense of continuous action. Along with such montage, *Blue Planet II* innovated with a camera able to join air and water into one well-focused image. This camera is the "megadome," a "24-inch dome lens that sits in front of the camera, which enables you to film and focus both above and beneath the water surface," in the words of Brownlow.[28] The filmmakers use the megadome in segments that expand the liquid fantasy of kinship with other marine mammals, and that further give this companionship additional emotional range beyond the strength and playfulness characterizing dolphins. Thus, the liquid fantasy of kinship incorporates apprehension and melancholy in the segment on the walrus mother and her pup that ends episode 1. The segment recounts the travails of a mother walrus in search of a resting place on an ice floe for her tired pup. She is repeatedly thwarted, as Arctic ice melts into the sea. Cast away in a warming polar ocean, the walrus mother and pup at once show the plight of their species and connect it to human climate refugees.

The liquid fantasy of kinship expands, on an optimistic note, to include the power of teamwork, in the segment on the Galápagos sea lions in episode 6, "Coasts." This segment shows how sea lions work together to hunt large, agile tuna who could easily outswim them in the open sea. Instead of chasing down the tuna, the sea lions herd them into the impasses of shallows rocks and run them up against the shore. Topside, underwater, and aerial shots give a panoramic view of this action, and again, the megadome creates unity between the atmospheres, as it enables a view of the porpoising sea lions on the hunt (fig. 4.4).

In such *Blue Planet II* segments modeled on fantasy, filmmakers convey the reality of never-before-seen behavior by fusing two kinds of vividness. The footage looks vivid as a result of camera and lighting technologies that yield imagery of exceptional clarity. The episodes feel vivid because we find our fantasies about the sea, in large measure created by film and television, realized in live action. Further, such an organization of marine reality along the lines of fantasy reinforces *Blue Planet II*'s message about the human connection to the sea—a concept so important to the series' message-making that its signature advertisement is composed as if it were photographed using the megadome, the camera whose shots epitomize the liquid fantasy of kinship. The top third of the image shows the walrus mother and pup, divided by a waterline from the bottom two-thirds of the image, which shows a sea lion in a kelp forest. Although these marine mammals only meet on screen, their conjoined image links diverse habitats, and spans water and air, to become an emblem of the ocean's connection to differing areas of our planet that should make humans care about ocean health.

FIG. 4.4. James Honeyborne and Mike Brownlow, producers, *Blue Planet II*, episode 6, "Coasts," ca. 9:03. Shot using the megadome, which can focus at once above and below water, in this view of a sea lion hunting tuna.

And yet, despite *Blue Planet II*'s efforts to create a sense of human connection with the oceans, its defining visual figures keep "distinctions between nature and society in place," as Wilson explains.[29] When humans appear, none of their roles defy *Blue Planet II*'s representations of the oceans as realms where we venture but we do not fundamentally belong. In episode 2, "The Deep," we see scientists and film crews exploring unknown depths, again to cite Wilson, "akin to the triumphant narratives of exploration from Europe's early frontiers."[30] When David Attenborough appears in the frame, he is the mild-mannered messenger of bad news about anthropogenic harm to wildlife. "Viewers of *Blue Planet II* were left shocked and distressed," according to an article in the *Evening Standard* the day after episode 4, where Attenborough steps on screen to dramatize a mother whale mourning her calf, likely killed by plastic in the mother's milk. The *Evening Standard* article also included a number of postings on social media in which appalled spectators vowed to abandon plastic.[31] Finally, humans figure as "pioneers who are striving to turn things around" in episode 7, "Our Blue Planet" (ca. 1:22). The episode features admirable scientists and conservationists working to offset human harm, with the aim of restoring damaged nature.

With *Blue Planet II*'s emotional appeal about the oceans as untouched wilderness, we can understand objections that the series was misleading viewers, when the *Guardian* reported that some crucial footage had been filmed in controlled settings, just as the series was about to premier.[32] The *Daily Mail* painted the use of studios in a sensationalist light, with its headline "BBC in new fakery row as it emerges viewers WON'T be told when scenes in flagship documentary Blue Planet II were filmed in laboratories rather than the wild."[33] The BBC responded by clarifying that constructed settings were used for filming the fangtooth in "The Deep"; for the coral bleaching event shown in "Coral Reefs"; for the zebra mantis shrimp in "The Green Seas"; and for the rock pools in "Coasts." There is nothing necessarily problematic about shooting documentary footage on a constructed set, if it enables the filmmaker to record real animal behavior that is hard to capture. Jean Painlevé and his colleague stayed up around the clock watching their aquarium so as not to miss the moment when the sea horse father expels the baby sea horses from his pouch once the eggs hatch, which they were the first to document on film. However, in a documentary fostering the view of nature as untouched by humans, the use of studio settings undercuts the myth.

In responding to the media flap, executive producer James Honeyborne declared, "You just can't break the spell," uttering a phrase he would repeat to the *Guardian*.[34] Yet Honeyborne also assured viewers that we want "a transparent relationship with our audience. . . . We want to tell all aspects of the oceans and we will use all the film craft techniques to do that."[35] Presumably in part to promote that transparent relation, the BBC produced short "making of" featurettes, which screened following the episodes. It also produced a ninety-minute follow-up, "Oceans of Wonder," showing how sequences were filmed. However, constructed settings were left out of the "making of" featurettes and the additional episode, suggesting that the BBC had a somewhat uneasy relation to their use. Honeyborne explained, "[W]e only have time for one making-of sequence at the end of each episode. We try to choose the one which was the most challenging to achieve, or the one in which the story of the crew's endeavours is most likely to engage the audience."[36]

The BBC's reluctance to display constructed settings also appears if we follow the links Honeyborne suggests for "the small proportion" of viewers interested in the ways in which views were captured, which, he specified, they "*can find out*."[37] That italicized phrase offers a hyperlink to a page with multiple tabs, including "Dive Deeper to See How Blue Planet Was Made."[38] The articles "Filming Close-up in Rock Pools" and "Filming Micro-Detail in the Deep" do describe the sets. Orla Doherty, for example, explains that camera people, dressed in "polar gear," achieved the fangtooth close-up "in a dark, refrigerated chamber on board the ship."[39] Nonetheless, there is not an image on the website of the camera people in "that dark cold room."[40] The use of a studio, moreover, is completely left off the page "Capturing the World's Bleaching Corals." Rather, the time-lapse footage is credited to a heroic scramble to document a coral bleaching event on the Great Barrier Reef, as it was occurring. The information portrays a cameraman, alerted to this devastation, who "set up underwater cameras on the reef in fixed positions, which then went on to film the tragic transformation for the next four weeks."[41]

One explanation for these moments when the BBC hesitates to reveal the role of human construction in *Blue Planet II* is that the series engages spectators by showing the oceans as pristine nature. This reluctance is echoed in the way Honeyborne justified studio filming in response to the *Guardian*, mentioning both "the welfare of the animals" or in other instances, "the safety of the scientists and crew."[42] Honeyborne did allude to the difficulty filming in marine environments—Richard Fleischer's "[t]he sea is there to defeat you"—in his justification for the rock pool scenes.[43] Thus, along with being "too disruptive for the wildlife," "it would have been impossible to film close-ups of this magical world, so we worked with scientists to accurately recreate a rock pool in the controlled conditions of the lab."[44] Honeyborne might have added to his "it would have been impossible to film close-ups of this magical world," the qualification, *as they conform to audience fantasies.* Indeed, the vividly colored anemones and other intertidal creatures filmed in Bamfield Marine Sciences Centre on Vancouver Island take us back even further than *Finding Nemo* to the chromolithographs by Philip Henry Gosse in *The Aquarium*, published a century and a half before. Relatedly, the deep-sea "fangtooth [which] has the largest teeth for its size of any fish" (ca. 13:55) continues a visual tradition for showing deep-sea anglerfish looming head-on, foreshortened, with a cold eye, massive jaw, and razor-edged teeth. This tradition goes back to the turn-of-the-twentieth-century images of the anglerfish in an account of the voyages of the German research vessel *Valdivia* and is still alive in *Finding Nemo*. *Blue Planet II's* episode "The Deep" also includes a glimpse of the anglerfish but surprises viewers when the anglerfish yields to the yet more dramatic fangtooth, "the midnight zone's most voracious fish" (ca. 14:15) (figs. 4.5, 4.6).[45]

The effectiveness of *Blue Planet II* is indisputable, even if there are debates about its degree. Nor can we challenge the reach of David Attenborough's eloquent pleas for environmental protection. Yet it is important to ask whether imagining the oceans without humans can limit as well as benefit environmentalist work. Such a vision is a fantasy as well, dating to Romanticism. As William Cronon writes so powerfully in "The Trouble with Wilderness; or, Getting

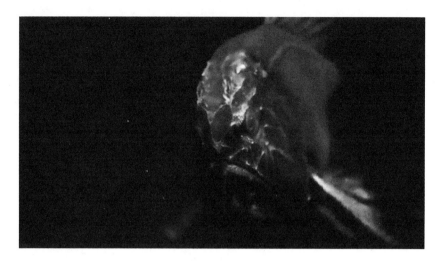

FIG. 4.5. James Honeyborne and Mike Brownlow, producers, *Blue Planet II*, episode 2, "The Deep" ca. 13:53. Fangtooth, "the midnight zone's most voracious fish."

FIG. 4.6. August Brauer, "Deep Sea Fish" (detail), in Carl Chun, *Wissenschaftliche Ergebnisse der Deutschen Tiefsee-Expedition auf dem Dampfer "Valdivia," 1898–1899* [*Scientific Results of the German Deep-Sea Expedition on the Steamer* Valdivia *1898–1899*] (Jena: J. Fischer, 1902–40), vol. 15, plate 15. From their first rendition, based on observation, deep-sea fish are imaged as horrifying creatures.

Back to the Wrong Nature": "*To the extent that we celebrate wilderness as the measure with which we judge civilization, we reproduce the dualism that sets humanity and nature at opposite poles. We thereby leave ourselves little hope of discovering what an ethical, sustainable, honorable human place in nature might actually look like*" (my italics).[46] In the past twenty years, Cronon's call to move beyond such dualism has been taken up by the environmental humanities and by posthumanist theorists such as Donna Haraway, Bruno Latour, and Timothy Morton.[47]

Humans are creatures who live in sustained interaction, whether directly or indirectly, with the oceans. How can we make a place for ourselves in reality and in imagination that goes beyond the dualism of intruders and heroes? In Wilson's discussion of *Blue Planet II*, she proposes the term "contact zone," which she adapts from postcolonial thinker Mary Louise Pratt to characterize, in Pratt's words, "social spaces where cultures meet, clash, and grapple with each other, often in contexts of highly asymmetrical relations of power."[48] Studying the off-screen film sets of *Blue Planet II*, Wilson expands this notion to encompass the materiality of physical environments, as well as the wide range of people who practice within them and whose competences enabled extraordinary imagery. On these film sets, "wildlife photographers, producers, technical crew, surfers, different communities, and researchers grapple with water, wind, rocks, and ice, plankton, plants, and all other manner of life across uneven and always shifting relations of power and knowledge-practices."[49]

Viewing the giant trevally and sooty tern sequence, for example, through the lens of the contact zone, we replace the liquid fantasy of apex predator carnage with a much messier if anthropologically gripping account. We also complicate the story of new technologies and first-world adventurers wresting unknown footage from the seas. The BBC first was alerted to such behavior by "a fisherman's tale that came to us out of South Africa," as executive producer Honeyborne said on a Royal Television broadcast about how the film was made (the trevally sequence does not feature on the BBC's "making of" featurettes, nor on the "Dive Deeper" website).[50] These reports were localized around the remote Indian Ocean atoll where *Blue Planet II* crews ventured. The islands were known to underwater cameraman Dan Beecham, who recounts that he had been based there for a decade during a previous job with the Save Our Seas Foundations.[51] To capture the giant trevally taking down a bird in midair, BBC crews required two trips to the Seychelles, in 2015 and 2016, involving also South African tour guide managers and Seychellois locals.[52] Along with the costly Phantom Flex camera, Ted Giffords depended on the water knowledge of Seychellois guide Peter King to capture the rapid giant trevally strikes that were beyond Giffords's ability to anticipate, no matter how fast his camera, eye, and hand. King, as Beecham explained, could "see the tiniest gesture the fish is making, even through the ripples on the surface of the water—he also has in [sic] innate understanding of the behaviour, that has taught him when the GTs will take an opportunity to grab a bird."[53]

These diverse interactions of humans with the oceans and with each other on location with *Blue Planet II* are a microcosm of environmental contact zones, which are objects of study for academics and pose challenges for policymakers. Could they also kindle the imagination of general audiences as the featured subject, rather than appearing in a "making of" featurette, an article, or on a sub-tab on a website for a "small proportion" of viewers? Along with awe-inspiring spectacles of wildlife, compelling imagery of how people in fact engage with the seas today is part of our great task to place humans justly in the frame.

NOTES

Introduction

1. See Marx, *History of Underwater Exploration*, for an overview of dive history, starting with the diving bells and free divers of antiquity and going forward to closed-helmet diving suits and other innovations.

2. See Cousteau and Malle, *Le monde du silence*. This print is shown at ca. 5:20.

3. Adamowsky emphasizes this point in *Mysterious Science*, 147–149. For the handful of scientific divers who experimented with helmet diving in the nineteenth century, see Heberlein, "Historical Development."

4. See Rozwadowski, *Fathoming the Ocean*, for a description and explanation of the emergence of this environment to public and scientific notice in the middle of the nineteenth century.

5. See Williams, *Triumph of Human Empire*.

6. See Corbin, *Lure of the Sea*.

7. On the invention of and enthusiasm for aquariums, see Brunner, *Ocean at Home*; Hamera, *Parlor Ponds*; and Adamowsky, *Mysterious Science*, among a number of works on the nineteenth-century installation of this spectacle.

8. *Literary Gazette*, July 15, 1854; the quotation in this paragraph is excerpted from an advertisement in Gosse, *Aquarium*, n.p.

9. Verne, *Twenty Thousand Leagues*, 109.

10. On Verne's mistakes, see marine biologist J. Malcolm Shick, "Otherworldly," 34. According to Shick, "as one goes deeper, not only is there less light to be seen, but long (red) wavelengths disappear first, followed by the progressively shorter orange, yellow and green. Blue light is left to penetrate deepest."

11. Adamowsky, *Mysterious Science*, 149.

12. On the impact of Williamson's photosphere, see Burgess, *Take Me under the Sea*; Crylen, "Cinematic Aquarium"; Adamowsky, *Mysterious Science*; and Elias, *Coral Empire*, among others.

13. "Mr. Hite Shows Underwater Pictures," 816.

14. "Submarine Photography," 479. The *New York Times* first mentions "submarine photography," as it was known until the 1950s, in 1932, when discussing Williamson's movies.

15. On the early modern *je ne sais quoi* in overseas travel literature, see Lamb, *Preserving the Self*.

16. Beebe, *Beneath Tropic Seas*, 6.

17. Wyllie's view of the octopus was perhaps shaped by Victor Hugo's well-known *The Toilers of the Sea* (1866), one of the first novels to portray a fisherman's thrilling open water struggle

with a monstrous octopus (*la pieuvre*). For Hugo, a demonic feature of the octopus was the way it killed by sucking out life.

18. Throughout, I translate the film title as *Sunken Ships*, rather than *Shipwrecks*, which is the English title given to the film. "Sunken Ships" is the translation used by Cousteau's English translator for "Epaves," the chapter on undersea wrecks in the 1953 *The Silent World*. *Naufrage* is another French word for "shipwreck," which suggests the process of sinking, in contrast to *épave*, used for "wreckage."

19. Shakespeare, *Tempest*, act 1, scene 2, lines 474–79.

20. Crylen, "Cinematic Aquarium."

21. Starosielski discusses new technologies in "Beyond Fluidity," as does Franziska Torma in "Frontiers of Visibility." See also Crylen, "Cinematic Aquarium," and Rozwadowski, "Arthur C. Clarke."

22. Malle, *Malle on Malle*, 8.

23. Young, *Making of 20,000 Leagues*, ca. 33:10.

24. On the technologies created for this capture, see Parisi, "Lunch on the Deck of the Titanic."

25. Keegan, *Futurist*, 174.

26. Crylen, "Cinematic Aquarium," 14.

27. As Shin Yamashiro observes, "Submarine scenery is in many ways characterized by a degree of terrestrial aesthetics" (*American Sea Literature*, 112).

28. "Global Shark Conservation." The Pew Charitable Trust estimates that "each year, at least 63 million and as many as 273 million sharks are killed in the world's commercial fisheries, many solely for their fins, which are used in shark fin soup."

29. Interviewed about the 1999 film in 2001, Howard Hall says that when making the film, "I was more worried about getting hit by a boat propeller or getting lost in a rainstorm" than about a shark attack. Fuchs, "Omnimax Film Attacks Shark Myths."

30. "Lloyd Bridges' Big Splash," 19.

31. Ibid.

32. See, for example, Rozwadowski, "Arthur C. Clarke," and Rozwadowski, "Ocean's Depths."

33. Alaimo, "Violet-Black." She is currently writing a book on the imagination of the deep sea in science, the arts, and popular culture whose working title is "Composing Blue Ecologies: Science, Aesthetics, and the Creatures of the Abyss."

34. Starosielski, "Beyond Fluidity," 152, 156, 159.

35. Ibid., 150–51. On marine mammals, see Mitman, *Reel Nature*, and Bousé, *Wildlife Films*. On sharks, see Ferguson, "Submerged Realities," and Benton, "Shark Films." While Sean Cubitt's chapter on *The Blue Planet* in *Ecomedia* focuses on the philosophical significance of the undersea, the book offers information useful for my questions in passing, such as Cubitt's explanation about the lighting system designed by James Cameron to permit viewing and filming the *Titanic* wreckage.

36. See, for example, Torma, "Frontiers of Visibility." Torma also notes that "interdisciplinary research on media, mobility and exploration has mainly covered land-based travel from the nineteenth to the twentieth centuries" (26).

37. D'Aloia, "Film in Depth."

38. See Mallinckrodt, "Exploring Underwater Worlds."

39. Walcott, "The Schooner *Flight*."

40. The file on this case is available on Global Shark Attack File, reference GSAF 1952.04.00, date: Sunday, April 1952, location: Suakin (Sawākin) Harbor, Sudan, accessed October 2020, http://sharkattackfile.net/spreadsheets/pdf_directory/1952.04.00-Hass.

41. Walcott, "The Schooner *Flight*."

42. DeLoughrey, "Submarine Futures," 40. Gender binaries are another example of categories undone in the marine environment, exemplified by the aesthetics of queer sea creatures in the painting of John Singer Sargent analyzed by Syme in *A Touch of Blossom* or the queer imaginings in the films of Jean Painlevé, noted by Erickson in "Animal Attraction, " and most recently the subject of Cahill's *Zoological Surrealism*.

43. Winkiel, "Hydro-criticism," 9.

44. This point runs throughout their writings. See Alaimo's previously cited articles and Mentz, for example, on the saltiness of the sea at the opening to Mentz, *At the Bottom of Shakespeare's Ocean*.

45. Cohen and Quigley, *Aesthetics of the Undersea*. See the essays in this volume for explorations of submarine materiality from the early modern period in the West. See also the essays in Abberley, *Underwater Worlds*.

Part 1: Vision Immersed, 1840–1953

1. Sylph, "Fish House."

2. Gosse, *Aquarium*, 171.

3. Adamowsky, *Mysterious Science*, 121. She cites, in English translation, Robert Geißler, *Plaudereien über Paris und die Weltausstellung* (Berlin: Theobald Grieben, 1868), 78–79.

4. Adamowsky, *Mysterious Science*, 121.

5. On both the continuity of aquariums with the aesthetic of collecting in the cabinet of curiosities, and their transformation of its displays, see Brunner, *Ocean at Home*, Hamera, *Parlor Ponds*, and Adamowsky, *Mysterious Science*, among others.

6. Adamowsky draws attention to the importance of the grotto for the aesthetics of the aquarium on 108–110. She describes the "three great aquariums made by Caumes and Bétancourt in the *jardin réservé*" for the Paris World's Fair of 1867 on 78: "Two of them had the form of grottoes. . . . The third . . . was built like a gigantic crystal chamber with walls and ceiling made of glass." A contemporary description of the Fair's aquariums is available at Worldsfairs.info, accessed October 2020, https://www.worldfairs.info/expopavillondetails.php?expo_id=3&pavillon_id=3788. On the early modern grotto's imagination of the submarine realm, see Rodríguez-Rincón, "Aesthetics of the Grotto," in Cohen and Quigley, *Aesthetics of the Undersea*.

7. Quigley, "The Porcellanous Ocean," in Cohen and Quigley, *Aesthetics of the Undersea*, 31.

8. Gosse, *Aquarium*, 171.

9. Ibid., vii.

10. Ibid.

11. Henri Milne-Edwards published his observations in "Rapport adressé à M. le ministre."

12. Adamowsky, *Mysterious Science*, 149.

13. Marx, *History of Underwater Exploration*, 18. Oviedo was observing the pearl fisheries on Margarita Island, where the Spaniards exploited Indigenous divers. As Marx observes, "many died from diseases brought from Europe by the Spaniards, while others died from exhaustion . . . being forced to dive as many as 16 hours a day."

14. I take this term from Ilardo et al., "Physiological and Genetic Adaptions."

15. Marx, *History of Underwater Exploration*, 49.

16. Adamowsky, *Mysterious Science*, 147–48.

17. For evidence of this challenge in depicting extremely remote environments, consider conspiracy theories that the Moon Landing was actually filmed on earth, perhaps by Stanley Kubrick, for example.

18. Milne-Edwards, "Rapport adressé à M. le ministre," 6. My translation; the French is "je voyais parfaitement tout ce dont j'étais entouré."

19. Marx, *History of Underwater Exploration*, 61.

20. Dickens, "Night Walks," 349.

21. Stevenson, "Random Memories." All Stevenson citations in this paragraph are from this text. Surveying popular Victorian works on coastal naturalism, I came across Sidney Dyer's *Ocean Gardens and Palaces, or, The Tent on the Beach*, published in 1880, which takes children along to witness the work of professional divers who describe for them the hazards of the depths, the variety of underwater work, and the beauty of coral reefs. At the same time, this book embellishes knowledge from divers with fantasy from aquariums or perhaps Verne, when, for example, it describes underwater vegetation as "blending in brightest hues all the colors of the rainbow" (79).

22. "Author—R. M. Ballantyne."

23. Ballantyne, *Under the Waves*, ii.

24. Ibid.

25. Ibid., 40.

26. Ibid., 82.

27. Ibid., 39.

28. Ibid.

29. On the history of Davis's publications, see the entry on Robert H. Davis on the website Classic Dive Books, accessed October 2020, http://classicdivebooks.customer.netspace.net.au/oeclassics-a-davis.html. The earliest edition of *Deep Diving and Submarine Operations* currently available at a library, according to Worldcat, is the fourth edition, printed in 1935.

30. Davis, *Deep Diving and Submarine Operations*. The description of the failed salvage operations for the HMS *Mary Rose* starts on 401. Seventeenth-century narratives include the use of a diving bell to recover treasure from a Spanish galleon sunk in Tobermory Bay, "a straggler from the rout of the Spanish Armada (A.D. 1588)," 337.

31. From *Punch* (1843), cited in Davis, *Deep Diving and Submarine Operations*, 337.

32. Quotes about the salvage of the *Eurydice* are from Davis, *Deep Diving and Submarine Operations*, 409–10. Davis further reports that the operation had to be delayed a number of times because of bad weather, when all the ships and divers returned to Portsmouth. After such challenges, the salvage eventually succeeded, and the next step was to "seal *Eurydice*'s ports preparatory to pumping her out."

33. Ransonnet-Villez, *Reise von Kairo nach Tor*, 18–19, cited in Adamowsky, *Mysterious Science*, 145, Adamowsky's translation. Ransonnet-Villez goes on to note seeing above him the "dunkler Schatten die Stelle des Boots," and hearing "das Geräusch der Wellen an das Ohr." On Ransonnet's interest in underwater drawing and painting, see also Jovanovic-Kruspel, Pisani, and Hantschk, "'Under Water.'"

34. Quotes from Ransonnet-Villez, *Ceylon, Skizzen seiner Bewohner, seines Thier- und Pflan-zenlebens und Untersuchungen des Meeresgrundes nahe der Küste* (Braunschweig: G. Westermann, 1868), 15–16, citation and translation in Jovanovic-Kruspel, Pisani, and Hantschk, "'Under Water,'" 141–42.

35. Adamowsky, *Mysterious Science*, 150.

36. See Heberlein, "Historical Development."

37. Haeckel, "Professor Haeckel in Ceylon," 389.

38. Fol, "Les impressions d'un scaphandrier." My translation. Fol first gave this talk to the leisured upper-class audience at the Club Nautique of Nice. He also wrote up his observations for a scientific audience published as "Observations sur la vision sous-marine faites dans la Méditerranée à l'aide du scaphandre," *Comptes Rendues Hebdomadaires des Séances de l'Académie des Sciences*, 1890.

39. Deane, "Underwater Photography," 1:1416. Thompson's image is reproduced in Gallant, "First Underwater Photo."

40. "Juvenile Books."

41. Ecott, *Neutral Buoyancy*, 33.

42. Wells, "In the Abyss," 81–82.

43. See Boutan, "Emploi du scaphandre."

44. Heberlein, "Historical Development," 288.

45. Taves, "Pioneer under the Sea."

46. For a succinct biography of Zarh Pritchard, see Shor, "Zarh H. Pritchard." I thank Eliza-beth Shor for generously sharing with me her detailed research on Pritchard. Details on Pritchard's life are also in Moure, *The World of Zarh Pritchard*, and the chapter on Pritchard in Burgess, *Take Me under the Sea*.

47. Vaughan, "Painting Beauty," 113.

48. Blanchon, "Zarh Pritchard," 3.

49. Pritchard, quoted in Vaughan, "Painting Beauty," 113.

50. Malle, *Malle on Malle*, 8.

51. Hass paraphrased by Adamowsky, *Mysterious Science*, 149.

52. See my "Underwater Optics," where I first expressed the ideas I refine in this section.

53. Beebe, *Beneath Tropic Seas*, 38.

54. Ibid.

55. Ibid.

56. Ibid.

57. Eldredge, "Wet Paint," 122. Hence Eldredge's coupling of artists undersea with the writ-ings of Melville, whom he views as a great modern myth maker of water symbolism (at the same time, Melville was also an experienced seaman drawing on the reality of maritime practice in his literature).

58. Ibid., 117–18.

59. According to Luria and Kinney, divers make errors underwater because "an underwater object is usually viewed by light that is insufficient and that has been scattered and drastically changed in wavelength, and the optical image has been modified in size and position" ("Under-water Vision," 1455).

60. Edge, *Underwater Photographer*, 8.

61. Beebe, *Beneath Tropic Seas*, 38.

62. Ibid., 36.

63. Ross, "Mist, Murk, and Visual Perception," 660.

64. Ransonnet, *Ceylon, Skizzen*, 15–16, cited by Jovanovic-Kruspel, Pisani, and Hantschk, "'Under Water,'" 142.

65. Pritchard quoted in Vaughan, "Painting Beauty," 113.

66. Pritchard in "Conversation with Zarh H. Pritchard," 219.

67. Diolé, *Undersea Adventure*, 15.

68. Panofsky, *Perspective*, 57.

69. Panofsky, *Perspective*, 57–58.

70. Ransonnet-Villez, *Travels from Cairo to Tor to the Coral Reefs* (1863), cited by Jovanovic-Kruspel, Pisani, and Hantschk, "'Under Water,'" 139, their translation, my emphasis. Ransonnet continues to explain that they were omitted, "to get a characteristic of the region rather than a synoptic representation of the different genera."

71. Leonardo da Vinci, *Notebooks*, 133.

72. Ross, *Behaviour and Perception*, 50.

73. Quotations from Ransonnet, *Ceylon, Skizzen*, 15–16, citation and translation in Jovanovic-Kruspel, Pisani, and Hantschk, "'Under Water,'" 141–42.

74. Ibid., 133.

75. Ibid., 150.

76. Adamowsky, *Mysterious Science*, 149.

77. Ransonnet, "Preface," cited in Jovanovic-Kruspel, Pisani, and Hantschk, "'Under Water,'" 143; emphasis in the original.

78. Jovanovic-Kruspel, Pisani, and Hantschk, "'Under Water,'" 143.

79. The Musée Océanographique in Monaco holds a collection of underwater art, started by Albert I, including the paintings of Ransonnet and Pritchard. Mark Dion included highlights of its collection for the exhibition *Oceanomania*, subtitled *Souvenirs of Mysterious Seas, from the Expedition to the Aquarium*, held at the Nouveau Musée National de Monaco in 2011. In 2013–14, the Tate St. Ives held an exhibition surveying both the history and contemporary practice of art engaging the underwater environment called *Aquatopia*, curated by Alex Farquharson and Martin Clark. In 2016, the Museu Marítim de Barcelona hosted *Underwater Painting: 150 Years of History* (featuring Ransonnet; Pritchard; André Laban, who dove with Cousteau; and Alfonso Cruz), Metropolitan Barcelona, accessed September 2020, https://www.barcelona-metropolitan.com/events/museu-maritim-barcelona-exhibition-underwater-painting.

80. Pritchard, in "Conversation with Zarh H. Pritchard," 219.

81. As Thomas Burgess writes, in Pritchard's painting, "no attempt was ever made to sharpen or refine images that were naturally softened" (*Take Me under the Sea*, 137).

82. Shick, "Otherworldly," 34.

83. Cited in Vaughan, "Painting Beauty," 113.

84. I thank Peter Brueggeman for his erudition in helping me to understand the details of Pritchard's paintings and how they correspond to optics at depth. According to Brueggeman, "Ellen Scripps gave the paintings to SIO. The gift year/date is uncertain. The gift is noted in an art object file here in Scripps Archives by a handwritten reference to SIO director Vaughan's papers as saying he mentioned the paintings in the Aquarium in 1924. . . . Paperwork

on the paintings in the art object file here in Scripps Archives says it was gifted in 1910 with no attribution" (personal e-mail, February 22, 2012).

85. Beebe, *Beneath Tropic Seas*, 36.

86. According to the *Chicago Daily Tribune*, Pritchard designed the costume she was to wear for Wilde's *Salomé*. "Bernhardt in Salomé: Gorgeous Costumes She Will Wear in Oscar Wilde's One-Act Play," *Chicago Daily Tribune*, August 28, 1892, 26.

87. Pritchard in "Conversation with Zarh H. Pritchard," 219.

88. Philip Hastings, e-mail to Elizabeth Shor, January 12, 2012. I thank Philip Hastings for his help in understanding the painting and also Elizabeth Shor for putting us in touch and sharing Professor Hastings's comments with me.

89. See, for example, Stott, *Darwin and the Barnacle*.

90. Philip Hastings, e-mail to Elizabeth Shor, January 12, 2012.

91. Panofsky, *Perspective*, 49. Panofsky contrasted the modern practice of linear perspective with earlier perspectival systems, such as that of the ancients, where bodies—substantial materiality—rather than space were the anchor of representation.

92. Ibid., 71.

93. See Jay, *Downcast Eyes*.

94. DeLoughrey, "Submarine Futures," 39.

95. DeCaires Taylor's website provides information about his diverse projects: accessed October 2020, https://www.underwatersculpture.com/?doing_wp_cron=1549147537 .2249300479888916015625.

96. Alaimo, "States of Suspension," 476.

97. Beebe, *Beneath Tropic Seas*, 41.

98. Jiro Harada, letter to Zarh Pritchard, San Francisco, July 17, 1915, in *Exhibition of Undersea Paintings by Zarh Pritchard*, Grace Nicholson Galleries (Pasadena, CA: Earl C. Tripp, 1926), 26.

99. All subsequent quotations of Pritchard in this paragraph are from "Conversation with Zarh H. Pritchard," 218, 219.

100. Newman, "Dream of Gerontius."

101. Burgess, *Take Me under the Sea*, 186. Factual details on Williamson's first film are in this paragraph from Burgess. A number of Williamson's films have been lost. The Library of Congress holds some stills. See Taves, "Pioneer." For reception of Williamson's first films, see also Bowers, *Terrors*. The Thanhouser.org website, by the corporation that subsequently purchased the film, credits Carl Louis Gregory as the director.

102. Crylen, "Cinematic Aquarium," 28n9 (note starts on 27).

103. Christie, "Set Design," 847.

104. "Visiting Goldfish" aka boblipton, comment thread, "Divers at Work on the Wreck of the 'Maine,'" accessed October 2020, https://www.imdb.com/title/tt0224357.

105. Burgess, *Take Me under the Sea*, 186.

106. Ibid.

107. Ibid.

108. Adamowsky, *Mysterious Science*, 160.

109. "In the Tropical Seas," Vimeo, accessed October 2020, https://vimeo.com/21137464.

110. Burgess, *Take Me under the Sea*, 210.

111. Ibid., 212.

112. Ibid., 211.

113. Ibid., 213.

114. George N. Shorey, review of *Twenty Thousand Leagues*, *Motion Picture News*, January 6, 1917, 112, cited in ibid., 212.

115. Gunning, "Cinema of Attraction[s]," 382.

116. Crylen, "Cinematic Aquarium," 45.

117. Gunning, "Cinema of Attraction[s]," 382.

118. Straten, "Annette Kellerman."

119. Williams had already made a film with the same title as Kellerman's first aquatic dance film shot in Bermuda in 1913, *Neptune's Daughter* (1949), although Williams would not perform underwater ballet until her subsequent *Jupiter's Darling* (1955).

120. Straten, "Annette Kellerman."

121. Straten, "Annette Kellerman." See also James Sullivan, *Venus of the South Seas* (1924), National Film and Sound Archive of Australia, accessed October 2020, https://www.nfsa.gov.au/collection/curated/venus-south-seas.

122. Slide, *Encyclopedia of Vaudeville*, 286.

123. Kenigsberg, "Tom Cruises's Most Dangerous Stunts."

124. Quoted in O'Hanlan, "Shipwreck of Reason," 140.

125. Erickson, "Animal Attraction." On Painlevé, see also Bellows, McDougall, and Berg, *Science Is Fiction*; Fretz, "Surréalisme sous-l'eau"; sections on surrealism and the scientific films of Painlevé in Adamowsky, *Mysterious Science*, such as "La Pieuvre—Jean Painlevé"; and Alex Zivkovic, "Underwater Embraces: Queer Ecologies in Jean Painlevé's Cinematic Science," unpublished honors thesis, Department of Art and Art History, Stanford University, 2017.

126. Painlevé, "The Seahorse," in Bellows, McDougall, and Berg, *Science Is Fiction*, xiii.

127. Breton, *Surrealism and Painting* (1928), quoted in Elias, "Sea of Dreams," 5. On Breton's interest in submarine materiality, see also my "Underwater Optics," and O'Hanlan, "Shipwreck of Reason." As O'Hanlan describes, Salvador Dalí too was interested in the underwater realm as figuring access to unconscious desires, and he appeared in a helmet diving suit at the International Surrealist Exhibition in London in 1936, although after ten minutes he was asphyxiating and had to remove the helmet.

128. O'Hanlan, "Shipwreck of Reason," 140.

129. Elias, "Sea of Dreams."

130. My translation. French: "Dans *Le Sang d'un poète*, j'essaie de tourner la poésie comme les frères Williamson tournent le fond de la mer. Il s'agissait de descendre en moi-même la cloche qu'ils descendent dans la mer, à de grandes profondeurs" (Jean Cocteau, quoted in Béhar, *Le cinéma des surréalistes*, 81).

131. Breton, *Mad Love*, 19.

132. Ibid., 11.

133. Ibid., 6.

134. Ibid., 13.

135. Georges Bataille, quoted in Krauss, "Corpus Delicti," 64.

136. Breton, *Mad Love*, 62.

137. http://www.lineature.com/en/motion/123-rogi-andre-photography-2.html. Accessed May 2014 (now inactive).

138. Magrini contextualizes this film in relation to the surrealist interest in breaking down categorical binaries in "'Surrealism' and the Omnipotence of Cinema."

139. Vigo, "Towards a Social Cinema."

140. Sitney, *Modernist Montage*, 28, 32.

141. Ibid., 32.

142. Man Ray, quoted in Béhar, *Le cinéma des surréalistes*, 156. In French, "effet de verre brouillé" using "quelques tranches de gélatine par impregnation."

143. Sitney, *Modernist Montage*, 32. He writes, "Such distortion as a sign of subjectivity had been part of the French cinema since Abel Gance's *La Folie de Dr. Tube* (1919)." This self-consciousness is both of the subject making the film and of the conditions for its production.

144. Ibid., 29.

145. Ibid., 32.

146. Aron, "Films of Revolt," 433.

147. I take the phrase "aquatic dreamworld" from Dolbear, "Flooded Displays," 4.

148. Aron, "Films of Revolt," 435.

149. Vigo, "Towards a Social Cinema."

150. Ibid. On Vigo's turn from surrealism to a more humanist cinema and his social aspirations, see Wild, "For a Concrete Aesthetics."

151. My translation. In French, "l'eau est son domaine comme celui du poisson."

152. Vigo, letter to Painlevé, Nice, May 28, 1931, quoted in "Film notes" about *Taris*, Il Cinema Ritrovato, accessed October 2020, https://festival.ilcinemaritrovato.it/en/film/la-natation-par -jean-taris-champion-de-france.

153. My translations. "Je n'ai rien vu." "Tu vas le voir un jour quand tu feras sérieusement."

154. "Sixteen Fathoms Deep," 113.

155. Pitts, *Columbia Pictures: Horror*, 16.

156. Gallant, "John D. Craig."

157. This information is on Worldcat.

158. Weiss, "Killers of the Sea," 23.

159. "16 Fathoms Deep," *Variety*, 12. "16 Fathoms Deep," *Times of India*, 9.

160. Schallert, "'16 Fathoms Deep'" 21.

161. Barry, "Film on Sponge Fishers," A4.

162. "16 Fathoms Deep," *Times of India*, 9. *Variety*, used to understatement was more reserved: the "deepwater denizens such as sharks and giant clams" add menace, but the underwater shots are at the same time somewhat anemic. They "have beauty but don't match supposed 16 fathoms deep location of action" (12).

163. "16 Fathoms Deep," *Variety*, 12.

164. "New Film, Modern Theatre," 10.

165. "Little Known Navy Heroes," D4.

166. Raimondo-Souto, *Motion Picture Photography*, 254.

167. "Unsung Heroes," 8.

168. Dash, "The Frogmen," 26.

169. Daugherty, "Arch Obeler," 11.

170. "Unsung Heroes," 8.

171. See Gevinson, *American Film Institute Catalog: Within Our Gates*, 84.

172. Crowther, "Wide-Angle Lens in Water," 52.

173. Ibid.

174. "Beneath the Twelve-Mile Reef," *Variety*, 6.

175. Ray Milland, quoted in *"Reap the Wild Wind*: Production Notes" (Universal DVD).

176. *"Reap the Wild Wind*: Notes" (Turner Classic Movies).

177. Crowther, "Screen in Review," L27.

178. Stafford, "Reap the Wild Wind."

179. *"Reap the Wild Wind*: Notes" (Turner Classic Movies).

180. Crowther, "Screen in Review," L27.

181. Stafford, "Reap the Wild Wind."

182. Ibid.

183. Ray Milland, quoted in *"Reap the Wild Wind:* Production Notes" (Universal DVD).

184. "All-Time Top Film Grossers," 69.

185. *"Reap the Wild Wind*: Notes" (Turner Classic Movies).

Part 2: The Wet Camera, 1951–1961

1. "1956 Le monde du silence."

2. In 1957, *The Silent World* won an Oscar for best documentary.

3. My translations in citing this article throughout. The French is "[l]es beautés du film sont d'abord celles de la nature et autant donc vaudrait critiquer Dieu." Bazin, *"Monde du silence,"* 35. In a review written in 1956 of Clouzot's entry, *The Picasso Mystery (Le mystère Picasso)*, Bazin remarked that he would have voted for Clouzot if he had been on the jury. Bazin, "Un film bergsonien: *Le mystère Picasso."*

4. I echo Crylen's notion of "enabling technologies" in "Cinematic Aquarium," 14. On the emergence of these technologies during World War II and their role in revolutionizing underwater films, see Torma, "Frontiers of Visibility," notably 33.

5. On the history of CCOUBA, see Quick, "History of Closed Circuit Oxygen Underwater Breathing Apparatus." On Williamson's use of British military suits, see Williamson, *Twenty Years,* 120ff.

6. Parry and Tillman, *Scuba America*, loc. 433, Kindle.

7. A good overview, contemporary with the advent of new camera equipment, may be found in *Underwater Photography* (1954), by Henry Kendall and Hilbert Schenck. While it does not credit Hans Hass, the book otherwise distinguishes important people in the joint pioneering of underwater photography and film, along with oceanography: "Jacques Gadreau of Multiphoto, Paris; Dr. Harold Edgerton . . . M. Dimitri Rebikoff; James Dugan and Capt. Jacques-Yves Cousteau; Willard Bascom, Raymond McAllister, Harris B. Stewart, Jr., Conrad Limbaugh and Thomas A. Manar" of SIO; and David Owen of WHOI and Jerry Greenberg of Marine Enterprises, among a number of people thanked at the book's opening. The intense innovation in underwater photography is indicated by the proliferation of guides about it, starting around 1954. Some innovations are mentioned on the (fairly) reliable website, Seafriends (see Anthoni, "Underwater Photography").

8. The *New York Times* obituary for Hass quotes Hass commenting, "For Cousteau, there existed only Cousteau. He never acknowledged others or corrected the impression that he wasn't the first in diving or in underwater photography" (comment made to dive historian Tim

Ecott in Vitello, "Hans Hass," A18). Viewing films by Hass, moreover, one finds similar images showing unique qualities of the undersea that Cousteau will note as well around the same years. Thus, Hass shows how air bubbles expand as they float up from the depths into the lighter atmosphere in his black-and-white *Humans among Sharks* (*Menschen Unter Haien*) (1947/48), an observation Cousteau will make in both the book (1953) and film *The Silent World* (1956).

9. Michael Jung, director of the Hans Hass Institut für Submarine Forschung, explains the technologies developed by Hass on a website dedicated to rebreathing history (see Bech, "Hans Hass").

10. Hass's water-housing replaced, for example Jacques Gadreau's Ondiphot water-housing, developed around 1950. I found the information about Hass's water-housing on "Rolleimarin Underwater Camera Owned by Neville Collins Photographer," http://www.hnco.com.au/rolleimarin-underwater-camera-owned-by-neville-collins-photographer in 2018, although this link is now broken. A record of the Ondiphot remains in sales of photos taken by Gadreau, such as in a 2008 auction catalog from the Drouot auction house in Paris, accessed October 2020: http://catalogue.gazette-drouot.com/pdf/piasa/photo/26092008/Piasa260908.pdf?id=2246&cp=75.

11. As a scientific example, the pioneering deep-sea vessel *Trieste*, which reached the deepest point in the ocean for the first time in 1960, was built in Italy according to the design of Swiss engineer Auguste Piccard, who navigated with it in the Mediterranean, before the US navy bought the vessel in 1958.

12. Writing about the United States in the postwar era, historian Gary Kroll comments, "Like a great flood tide, the ocean rushed into the bays and coves of post-war American culture." He observes, "The explanation for why so many Americans took to the water defies simple explanation" (*America's Ocean Wilderness*, 96, 135). Kroll's observations are true of other countries in the vanguard of modernization, notably those with territorial access to warm water.

13. Cousteau and Dumas receive credits as coauthors in the French version, while the English edition says "with Dumas," and credits Cousteau as author. Details also differ in small but culturally meaningful ways. Thus, at the opening of the French version, Cousteau mentions his strong emotion and portrays loading the first aqualung from the train station onto "la fidèle voiture que Roger Gary abreuvait de dissolvant pour peintures," used as a gas substitute in wartime France (Cousteau and Dumas, *Le monde du silence*, 7). The English translation meanwhile clarifies the vague "regards indiscrets" that the divers seek to avoid as those of "curious bathers and Italian occupation troops" (*Le monde du silence*, 8; *The Silent World*, 4).

14. Ayer Directory of Publications. Cathy Hunter, archivist at *National Geographic*, estimates that "outside of the US, circulation was probably highest in Canada and the UK, with some also going to Europe, Scandinavia, and Mexico)." Victoria Googasian, e-mail to author, August 2, 2017.

15. British Broadcasting Corporation, "Audience Research Report," June 1, 1956.

16. Rozwadowski, "Arthur C. Clarke," 583. See Rozwadowski's *Fathoming the Ocean* for an account of how at the time the aquarium was invented in the nineteenth century, the turn of attention to the ocean frontier also involved the takeoff of oceanography and modern marine biology, connected in addition to the laying of transoceanic cables, innovations in marine engineering, and the first systematic survey of the ocean depths.

17. Rozwadowski, "Playing by—and on and under—the Sea," 184–85.

18. I found these films in part through threads on diver websites such as Divermag and Scubaboard.com, which I discovered thanks to Peter Brueggeman (see, for example, accessed

October 2020, http://divermag.com/the-best-diving-movies-of-all-time and the thread that follows). In *Creature from the Black Lagoon* (1954), costume designers tried to incorporate scuba tanks for the creature, but found them too bulky, so instead the actor who plays Gill-man, Ricou Browning, performed breath-holding, taking breaths from airhoses hidden on the set (Voger, "Browning Performed Swimmingly"). Browning would go on to play an underwater double in *20,000 Leagues under the Sea.* The proximity of Hollywood to Southern California dive and dive photography culture, which helped spearhead technical innovation, contributed both to the film industry's awareness of the underwater environment's potential as a film set and to its technical ability to set up shop there. Rozwadowski calls the Scripps Institution of Oceanography "ground zero for scuba" ("Playing by—and on and under—the Sea," 175). There are fewer references in this period to documentary films for general audience distribution, and it may be that fewer documentaries were made for general release.

19. Hanauer, *Diving Pioneers,* 35. *Sea Hunt* would become "the country's most successful first-run syndicated TV show at that time [until] *Baywatch* (1989–2001)," according to the now-archived website by popular scuba historian Bill Jones, http://home.comcast.net/~bill.jones .scubaguy/SEA_HUNT/SeaHunt.html. I accessed this website in 2015, although the link is now broken.

20. Diolé, *Undersea Adventure,* 178.

21. Bazin, "*Monde du silence,*" 37.

22. Malle, *Malle on Malle,* 9, 8.

23. Readers who would like to get a sense of the sophisticated shaping of underwater exhibition in documentaries by Hass, Cousteau, and Malle can contrast them with scientific films from the 1950s, such as those in the media archives of the Scripps Institution of Oceanography. The portal to its collection, accessed October 2019, is https://library.ucsd.edu/research-and -collections/collections/special-collections-and-archives/manuscripts/scripps-archives.html.

24. German: "Alle gezeigten Aufnahme wurden original am Roten Meer gedreht . . ." The timestamp is for the German version. The line does not appear in the English version released as *Under the Red Sea.*

25. Hanauer, *Diving Pioneers,* 153.

26. Black, "Teleview."

27. Bazin, "Le cinéma et l'exploration," quotes in this paragraph found on 31, 32. My translations.

28. Malle, *Malle on Malle,* 8.

29. Victoria Googasian offered this astute observation. Other articles about the arduous production of underwater films include Cousteau's article on difficulties shooting under the sea in the *New York Times* ("Studio under the Sea,"); the difficulties posed by "giant jellyfish . . . alarming creatures, trailing transparent tentacles up to ten feet long" in filming *The Silent Enemy* (Watts, "Activities," 5); and the perils of shooting sharks observed in Raines, "Movie Log," 7, on making *The Old Man and the Sea,* including underwater sequences by Lamar Boren. They also include *Life,* "Film Stars with Fins," and articles in *National Geographic,* such as Marden, "Camera under the Sea" and Edgerton and Cousteau's articles.

30. "Weird New Film World," 111.

31. *Making of 20,000 Leagues,* ca. 37:36.

32. Torma, "Frontiers of Visibility," 28.

33. On soundscapes in Cousteau, see Cooke, "Water Music." David Chapman discusses sound in Cousteau and Hass in "The Undersea World of the Sound Department."

34. The title of *The Silent World* is misleading. In the postwar era, it became common knowledge that marine organisms emitted sounds and, moreover, that other types of sound traveled under water. In the words of an 1947 article from the *Biological Bulletin*, "The sea has long been looked upon as a realm of silence. That this view is no longer tenable has been abundantly demonstrated . . . with modern underwater sound detecting gear. These investigations, which during World War II became of grave importance because of submarine warfare and harbor defense, have shown that great stretches of numerous coastal areas are exceedingly noisy, periodically or perpetually, due to marine animals" (Johnson, Everset, and Young, "Role of Snapping Shrimp," 122).

35. Mitman, *Reel Nature*, 175.

36. Compare the dramatic depiction of underwater space in *20,000 Leagues under the Sea*, for example, to the more diffuse exhibition in Robert Webb's *Beneath the 12-Mile Reef* (1953).

37. Two cameras were used for filming underwater: the Aquaflex and Hollywood's standard in air, the Mitchell camera, enclosed in "a watertight case." "*20,000 Leagues Under the Sea*," AFI Catalogue.

38. *Making of 20,000 Leagues*, ca. 43:18.

39. Condee, *Theatrical Space*, 66.

40. *Making of 20,000 Leagues*, ca. 32:30.

41. Tom Hutchinson, "20,000 Headaches," 12–13.

42. The name of this ship is *The Florida*, an unusual name for a British ship, as Butcher observes, noting that "it must be borrowed from the Confederate *Florida* which sank 37 Northern ships in just over a year" (Verne, *Twenty Thousand Leagues*, 408n125).

43. The timestamp for *Under the Red Sea* here and throughout refers to the English version of the film, although my analysis refers to details that are in some cases only found in the German version. The original German version is almost fifteen minutes longer, is edited differently, and includes sequences left out of the English version. Hass's narration in the German version has a more personal tone than the didactic narration of the English version spoken by Les Tremayne, and the German version includes dialogue among the members of the expedition. A webpage of the Film Archiv Austria offers a description of the German version: https://www.filmarchiv .at/en/program/film/abenteuer-im-roten-meer/.

44. I published early drafts of some of the material covered in this section in "The Underwater Imagination," 51–71; and "The Shipwreck as Undersea Gothic," 155–66.

45. Cousteau: "At night I had often had visions of flying by extending my arms as wings. Now I flew without wings. (Since that first aqualung flight, I have never had a dream of flying)" (Cousteau with Dumas, *Silent World*, 6).

46. Ibid.

47. Mercado, *Filmmaker's Eye*, 167.

48. Malle, *Malle on Malle*, 8.

49. Marden, "Camera under the Sea," 170.

50. Cousteau, *Window in the Sea*, 96.

51. Crylen, *Cinematic Aquarium*, 17 and throughout. Torma too notes this fantasy of what she calls "mermen" (a term in circulation throughout the era of the 1920s-1950s), shared by Hass as well. Torma, "Frontiers of Visibility," 33–34.

52. Cousteau, *Window in the Sea*, 96.

53. Ibid.

54. On Edgerton's underwater contributions, see Calcagno, "Underwater Photography."

55. Cousteau, *Window in the Sea*, 60. We would have been shown an underwater grotesque rather than an underwater glide, had Cousteau used his initial design for wet suits that made him look "something like Don Quixote" (Cousteau with Dumas, *Silent World*, 13).

56. Cousteau with Dumas, *Silent World*, 12.

57. Sidney, "Diving with 'Jupiter's Darling,'" X5.

58. Crylen observes of "Cousteau's roving undersea images" that the camera "move[s] with a fluidity unusual for handheld camera," stabilized by the water ("Cinematic Aquarium," 69).

59. Cousteau and his fellow divers were the first to understand that by depths of 140 feet—the exact threshold depends on the physiology of each diver—gases such as nitrogen in the air supply, usually inert at the pressure on land and released into the bloodstream of the diver, cause the feeling of a high that impairs judgment. Cousteau termed this feeling *l'ivresse des grandes profondeurs*. *Ivresse* is usually translated usually as "rapture," but a more literal translation is "intoxication." In a colloquial American metric, this intoxication is called the martini effect. The deeper the diver goes, the more intense the effect: divers have estimated that each fifteen meters of depth corresponds to the impact of having drunk one martini.

60. Malle, *Malle on Malle*, 9.

61. Adrienne Rich captures the topside mood of optimism generally prevailing on the deck of the *Calypso* in her poem, "Diving into the Wreck," when she contrasts her own dive alone to "Cousteau with his / assiduous team / aboard the sun-flooded schooner" (the *Calypso* is not a schooner; this is a poetically meaningful substitution, as schooners are swift and maneuverable sailing crafts, historically used for smuggling, among other uses).

62. Here and in the next two paragraphs, I quote features of film noir from McDonnell, "Film Noir Style," 73–74.

63. This and subsequent quotations from the narration are my translations of the French version. Here, the French is "les cerveaux sont guettés par l'ivresse des profondeurs."

64. Cousteau, "Fish Men," 439.

65. The French is: "Plus ils descendent, plus les bruns deviennent chauds, plus les rouges deviennent violents."

66. Marden, "Camera under the Sea," 196.

67. Cousteau with Dumas, *Silent World*, 245.

68. Ibid., 247.

69. Cousteau with Dumas, *Silent World*, 209.

70. The French is: "Il ne faut pas aller plus bas. Il ne faut pas rester longtemps."

71. The French is: "groupes d'organismes de formes étranges" (ca. 18:23); "un ciel d'étoiles" (ca. 18:18).

72. Cousteau with Dumas, *Silent World*, 41.

73. "Wreck Diving: Wrecks of the Red Sea; Thistlegorm."

74. "The Thistlegorm Project."

75. Radcliffe, *Mysteries of Udolpho*, phrases from 326–28.

76. Mercado, *Filmmaker's Eye*, 101. All quotations about the canted shot are from this page.

77. Cousteau with Dumas, *Silent World*, 40.

78. The soundtrack of *Sunken Ships* is more descriptive when the diver explores a wreck, as the voiceover narration evokes ghosts, dream, and danger. The narrator also observes the numerous fish there, calling the wreck an "oasis" (ca. 12:58), in a "desert of sand" (ca. 12:54).

79. Bazin viewed the exhibition of the dead in a purely informational manner without "any moral or aesthetic justification" as obscene, remarking that in a 1956 program on the *Journal télévisé*, which had underwater sequences "rather in the style of *Le Monde du silence*" the drowned countenance of a pilot was treated "as if it were just another curiosity" (Bazin, "Information or Necrophagy," 124). I thank Dudley Andrew for drawing my attention to this passage.

80. The French is "un vieux timonier l'a pris en main avec orgueil pour son premier appareil-lage. Et ce tube c'était un canon. Le volant de pointage tourne encore."

81. Sorensen, "Ottawa Shows Off Bell." I thank Adriana Craciun for this insight.

82. Molesworth, "Gothic Time," 38. Molesworth writes, "The gothic clock acknowledges the temporal modernity described by Thompson and others through its sheer ubiquity, but the time discipline it enforces owes less to the rise of industrial capital than to a strongly developed sense of ritual practice" (45).

83. Cousteau with Dumas, *Silent World*, 39.

84. The French is: "Tous les marins du monde détestent les requins."

85. The French is: "La ronde infernale des requins autour de la proie—si des hommes l'ont déjà vue, ils n'ont pas vécu pour le dire."

86. Torma, "Frontiers of Visibility," 37.

87. The French is: "imprudent comme tous les gosses"; "lutte courageusement pour rejoin-dre sa famille."

88. The French is: "mais la ruse est impuissante contre le dynamite."

89. The film poses this friendship as a question after showing companionship between fish of two different species: "l'amitié d'un homme et d'un poisson, serait-elle plus insolite?"

90. Marden, "Camera under the Sea," 165.

91. The French is: "Jojo se trouvait bien avec nous. Nous l'avons quitté avec regret."

92. Morris, "Sea-Bed."

93. "Down to the sea bed for adventure," *Birmingham Mail*. Other films mentioned in this article include Peter Scott's *Look* and "the jungle pictures of Michaela and Armand Denis." A month later, Winifred Carr noted that Lotte now had competition from Ramona Morris, who appeared with her husband Desmond in "Zoo Time," first broadcast on May 8, 1956 (after five of the six episodes of *Diving to Adventure*). David Attenborough was making a name for himself as well in this first era of TV documentaries featuring wildlife, with episodes in the series *Travel-ers' Tales* (1956–57) and *Zoo Quest* (1954–61).

94. "Hans Hass," http://www.wildfilmhistory.org/person/102/Hans+Hass.html. In January 2019, the Wild Film History website was taken offline. According to the website, "The collection is stored securely offline in perpetuity." For information, contact hello@wildscreen.org.

95. Derek Bousé does not identify any underwater TV films, let alone series, before *Diving to Adventure*. See Bousé, *Wildlife Films*, 214. Other early underwater TV films include *Mysteries of the Deep* (1959), a short that is part of Disney's True-Life Adventures series. and Hass's sub-sequent 1958 BBC series *The Undersea World of Adventure*. *The Undersea World of Jacques Cous-teau*, a twelve-film series, had a pilot in 1966, and then ran with regularity starting in 1968.

96. "*Diving to Adventure: In the Aegean* (1956), http://www.wildfilmhistory.org/film/80 /In+the+Aegean.html (now inactive). See note 94 for contact information to access the collection.

97. BBC Audience Research Department, "Report." The report rated their popularity behind previous Cousteau talks in 1950 and 1953, which "reached even higher Indices 84 and 86."

98. Torma, "Frontiers of Visibility," 41.

99. Norton, *Underwater*, 6.

100. "BBC View," *Birmingham Mail*.

101. Crowther, "Screen: Beautiful Sea," 30.

102. "Dangers of Underwater Swimming."

103. Ibid.

104. I thank Michael Jung of the Hans Hass Institut für Submarine Forschung und Tauchtechnik for the archival material about Hass in this chapter. I also thank Reg Vallintine of the Historical Diving Society for recordings of *Diving to Adventure*.

105. North, "Battle of the Beaches."

106. Gugen, letter to Hans Hass, London, April 6, 1956. Archive, Hans Hass Institut für Submarine Forschung und Tauchtechnik.

107. This phrase is from Pound, "Critic on the Hearth," 424.

108. Crowther, "'Under the Red Sea,'" 37.

109. Ibid.

110. Marsland Gander, *The Daily Telegraph*, 1936, quoted in "Here's Looking at You."

111. Torma, "Frontiers of Visibility," 31.

112. Crowther was tart about the female star when she was not integrated into the action, hence his comments about *Underwater*, in which Jane Russell plays an underwater treasure hunter: "Even Miss Russell, heavily laden in a tangle of skin-diver's gear, is no one to excite much interest, let alone raise the temperature. . . . Miss Russell does nothing underwater that she hasn't done better above" (Crowther, "'Underwater!,'" L27).

113. The attraction of the bathing beauty went underwater in the theme park at Weeki Wachee, Florida, following World War II. The park opened on October 13, 1947, using "a method of breathing underwater from a free-flowing air hose supplying oxygen from an air compressor, rather than from a tank strapped to the back. With the air hose, humans could give the appearance of thriving twenty feet underwater with no breathing apparatus." In fact, "mermaids performed synchronized ballet moves underwater while breathing through the air hoses hidden in the scenery" ("Weeki Wachee's History").

114. Fox, "Lotte Hass, Pioneering Diver," D7.

115. Barrett, "Lotte Is Going Down Again," 12.

116. The first female astronaut was Russian Valentina Tereshkova for the Soviet space program in 1963, and the United States would not send a woman into outer space until Sally Ride in 1983.

117. Torma, "Frontiers of Visibility," 31.

118. Crosby, "Cinematographer Dr. Hans Hass," 8.

119. "The Sublime, the Picturesque, the Beautiful," guide to the exhibition *American Scenery* (2012), accessed December 2018, https://blantonmuseum.org/files/american_scenery /sublime_guide.pdf. This guide has now been taken down by the Blanton Museum, but I cite it because it distills in three pages foundational observations from the aesthetic theories of Enlightenment/pre-Romantic British aesthetic theorists, William Gilpin, Uvedale Price, and Edmund Burke.

120. See Elias, *Coral Empire*, throughout for an analysis of the changing representation of coral in twentieth-century Western modernity.

121. Diolé, *Undersea Adventure*, 134. See Schiller, "Der Taucher" [The Diver] (1797). Edward Bulwer-Lytton's translation, accessed December 2020, is available at http://www.bartleby.com/360/9/10.html.

122. [Untitled on *Diving to Adventure*], *Birmingham Mail*.

123. Burke, *Philosophical Enquiry*, 48.

124. Pound, "Critic on the Hearth," 424.

125. Watt, "Deep Sea 'Ballet.'"

126. Pepys, "Underwater Grace."

127. Ibid.

128. Morris, "TV Remembers the Divers," 3.

129. Crowther, "'Under the Red Sea.'"

130. This sequence appears in the original German version of *Under the Red Sea* starting at ca. 1:08:18.

131. [Untitled on *Diving to Adventure*], *Birmingham Mail*.

132. Crosby, "Cinematographer Dr. Hass," 8. The appreciation may have been specific to those familiar with the underwater environment. *Variety* was less enthusiastic, remarking that "[t]his sequence . . . and much of the narrative have a 'fishy' feel" ("Under the Red Sea," *Variety*, 22).

133. "Lloyd Bridges' Big Splash," 17.

134. Ibid.

135. Ibid., 18. According to an article about the success of *Sea Hunt* at the beginning of its third season, "Tors finds more fun in TV production because an idea is translated onto the screen so much quicker than in motion pictures. A feature is nursed for years, at times, before the final product is realized. In telefilms, once the initial plunge is made, the span is much shorter" ("Tors Envisions $1,000,000 Take," 30).

136. *Official Guide, New York World's Fair 1964/1965*.

137. Accessed September 2015, http://home.comcast.net/~bill.jones.scubaguy/SEA_HUNT/SeaHunt.html (now inactive). Jones noted that the TV series "aired in at least 20 other countries in eight languages other than English," including "Australia, Cuba, France, Germany, Japan, Mexico, the Philippines, Puerto Rico, the United Kingdom, the Virgin Islands, and the list goes on. In April 1958, SEA HUNT was dubbed in Russian and was included as a part of the first swap of TV programs under a U.S. and Soviet cultural exchange program."

138. *Sea Hunt*, "The Sea Has Ears," September 27, 1958, season 1, episode 38, ca. 25:00.

139. See my *The Novel and the Sea* for a discussion of this pattern.

140. Joseph Conrad, "Author's Note to *Typhoon*," 218.

141. On Boren's contribution to underwater film, see Hanauer, *Diving Pioneers*, 49–50.

142. "Using exceptionally fast film (Dupont Super 4, which is 13 times faster than ordinary black-and-white), Tors continued, 'we know anything we can see with our own eyes, the camera can see better.'" Both of these quotations are from "Even the Secretary Goes under Water," 27. The secretary was played by Parry.

143. "Light is adequate at most depths, particularly at a sandy bottom, which reflects light rays" ("Even the Secretary Goes under Water," 27).

144. "Lloyd Bridges Aqualung Actor," 23.

145. "Lloyd Bridges' Big Splash," 19.

146. "Tors Envisions $1,000,000 Take from 'Sea Hunt,'" 30.

147. "The Raft," *Sea Hunt*, episode identification 72B, 34.

148. An early version of the argument in this section appears in my "Adventures in Toxic Atmosphere."

149. According to an Internet database dedicated to identifying aircraft in movies and TV shows, "On 4 November 1954, Convair YF2Y-1 Sea Dart, BuNo 135762, disintegrated in midair over San Diego Bay killing Charles E. Richbourg." In "Sixty Feet Below," the pilot "is just prisonner [*sic*] of the sunk aircraft (named XF-190). The experimental aircraft is played by XF2Y-1 prototype (BuNo 137634) with its original J-34 engines and twin ski (retracted here)." "Sea Hunt," IMPDb, http://www.impdb.org/index.php?title=Sea_Hunt.

150. Barthes, *S/Z*, 75.

151. In Barthes's astute, if technical, analysis, "the problem is to *maintain* the enigma in the initial void of its answer; whereas the sentences [agents of what he calls the code of actions] quicken the story's 'unfolding' and cannot help but move the story along, the hermeneutic code . . . must set up *delays* (obstacles, stoppages, deviations) in the flow of the discourse." Ibid.

152. Wulff, "Suspense and the Influence of Cataphora." *Cataphora* comes from ancient Greek and its etymology fortuitously suggests the act of carrying downwards.

153. On the use of clocks as cataphora in *High Noon*, see Powell, *Stop the Clocks*.

154. The exhibition of "dead time" as the time of lived, human experience, is prominent, in contrast, in independent cinema reacting against Hollywood conventions, notably in films of the French New Wave. Directors such as Jacques Rivette or Chantal Akerman, for example, bring screen and plot time together to show that daily life and its phenomenology are vastly more complex than the action pacing demanded by the entertainment industry.

155. Verne, *Twenty Thousand Leagues*, 320. Subsequent quotations are referenced by page number within the text.

156. "Lloyd Bridges' Big Splash," 18.

157. Parry quoted in Hanauer, *Diving Pioneers*, 153.

158. "Lloyd Bridges' Big Splash," 17–18.

Part 3: Liquid Fantasies, 1963–2004

1. See Dimmock and Cummins, "History of Scuba Diving Tourism."

2. Carter, "Media: TELEVISION; Stand Aside," 41.

3. See Crylen, "Cinematic Aquarium," for a discussion of the underwater habitat tradition. The science fiction *Voyage to the Bottom of the Sea* (1961) is an influential submersible film, followed by the eponymous TV series (1964–68), which includes diving in its adventures.

4. These facts are from the *Flipper* trivia page, IMDb.

5. Mitman, *Reel Nature*, 178. In Mitman's words, "[t]he making of the dolphin into a glamour species within American culture was, like the making of natural history film, the result of much behind-the-scenes labor in which scientific research and vernacular knowledge, education and entertainment, and authenticity and artifice were edited and integrated into the final scenes that appeared before the public" (177).

6. Ibid., 165, 166.

7. Ibid., 168.

8. Mitman further underscores the role of scientists and the military in this history, and *Flipper* thanks John Lilly for his "scientific advice" in the film's credits (ca. 1:29:48).

9. Burden cited in Mitman, *Reel Nature*, 160.

10. Mitman, *Reel Nature*, 160.

11. Quotes in this paragraph are from Shepard, "Flipper, the Educated Dolphin," 23.

12. Browning, paraphrased by Clarke, "James Bond Doesn't Do CGI."

13. Facts in these three sentences are from Chancellor, *James Bond*, 29, 43.

14. Ibid., 103.

15. Ibid., 43.

16. Browning, quoted by Clarke, "James Bond Doesn't Do CGI."

17. "James Bond Retrospective: Thunderball (1965)." *Goldfinger* (1964), directed by Guy Hamilton, was the Bond film before *Thunderball*, and the third one made.

18. Crowther, "Screen: 007's Underwater Adventures," 23.

19. Ibid.

20. Cork, *Behind the Scenes with "Thunderball*," ca. 20:35.

21. "Topics," *New York Times*, 26. Credit was given by the article to "J. Y. Cousteau and Emile Gagnan," with their Aqualung, as well as to Cousteau's book. In addition, the article noted that color photography (erroneously first credited to Cousteau) and "[m]otion pictures of the brilliantly colored denizens of the deep have astounded, attracted and intrigued mere earthlings."

22. Jackson, "10 Fascinating Facts."

23. Wilkinson, "Inside James Bond's Lotus Supersub."

24. Cork, *Behind the Scenes with "Thunderball*," ca. 18:37.

25. Ibid., ca. 21:47.

26. Ibid., ca. 22.11.

27. Binder "was a ceaseless experimenter—with colour (emulsions, filters), optical effects, underwater shots, slow motion, animation, electronic movement, superimposition and ways to blend or overlap images" (Kirkham, "Dots and Sickles," 12).

28. Johnson, "Sex, Snobbery, and Sadism."

29. "Exception Form: November 25, 1965. *Thunderball.*"

30. Ibid.

31. Ibid.

32. Mulvey, "Visual Pleasure," 838.

33. Fernbach, "Fetishization of Masculinity."

34. Mulvey, "Visual Pleasure," 837.

35. Sean Hutchinson, "25 Facts about *Jaws.*"

36. See, for example, Willson, "*Jaws* as Submarine Movie"; Dowling, "'Revenge upon a Dumb Brute'"; Rushing and Frentz, *Projecting the Shadow*.

37. Canby, "Screen: Dramatic Pursuit," 51.

38. Canby noted that "the language is not equal to the event" by quoting one of Gimbel's comments: "I got the liver scared out of me at one point." Ibid.

39. Ibid.

40. Ibid.

41. Kay was played by Julie Adams and doubled in the underwater scenes by Ginger Stanley—who also, according to IMDb, doubled in *Jupiter's Darling*.

42. Sweeney, "Jaws (1975)."

43. "On Location with *Jaws*." As the article specifies: "One of the interesting pieces of equipment that Butler developed for this marine movie was a special raft that could be raised or lowered (on its pontoons) out of the water to different levels. It also had a section cut out on one side that would fit the water box that was used to protect the Panaflex camera during shots that were done right on or in the water."

44. Gordon, *Empire of Dreams*, 35.

45. Ibid., 32.

46. Ibid., 34

47. Elias, "Frank Hurley," 124.

48. Ibid.

49. Carroll, *Philosophy of Horror*, 37.

50. Ellis, *Shark*, 88.

51. "Shark Week 2019 Announced."

52. Moody, *EMI Films*, 147.

53. Guber with Witus, *Inside The Deep*, 10.

54. Ibid.

55. Giddings, quoted in Gilliam, "Al Giddings, Dean."

56. Ibid.

57. Ibid.

58. "The Deep," *Variety*.

59. Griffin and Masters, *Hit and Run*, 86.

60. Moody, *EMI Films*, 148.

61. "Best Scuba Diving Movies of All Time."

62. "The Deep," *Variety*.

63. Canby, "'The Deep'" 10.

64. "The Deep Reviews," *TV Guide*.

65. Quoted in Griffin and Masters, *Hit and Run*, 86.

66. Guber with Witus, *Inside The Deep*, 10.

67. Rozwadowski, "Arthur C. Clarke," 596, 597.

68. Language from the trailer for *The Monster That Challenged the World*.

69. Quoted in Keegan, *Futurist*, 9.

70. Keegan, *Futurist*, 88.

71. Charles Nicklin, conversation with the author, La Jolla, CA, November 29, 2016.

72. "Biography: Pete Romano."

73. Keegan, *Futurist*, 93.

74. Ibid.

75. Frakes et al., *James Cameron's Story of Science Fiction*, 16.

76. Ibid., 65. Cameron lists this film among the top ten science fiction films in *James Cameron's Story of Science Fiction* and cites it repeatedly in *The Abyss*.

77. Don Vaughan, "Chesley Bonestell."

78. Milton cited by Burke, *Philosophical Enquiry*, 49.

79. Alfonso Cuarón dismantles this disturbing disconnect in *Gravity* when he turns the astronauts themselves into vertiginous, rotating bodies throughout all their time outside the craft, creating at least fatigue, if not motion sickness, for viewers.

80. Wells, "In the Abyss," 88.

81. Alaimo, "Dispersing Disaster," 185.

82. Alaimo, "Violet-Black," 238, 239.

83. Ibid., 245.

84. Keller, *James Cameron*, 117.

85. Kroll uses this term, and Alaimo cites it as well.

86. Keegan, *Futurist*, 85.

87. Ibid., 85.

88. Video available on Dillow, "First Footage from Cameron's Dive."

89. Quoted in Milman, "The Deep Ocean."

90. Quoted in Hayward, *Luc Besson*, 6; my translation.

91. Hayward, *Luc Besson*, 49.

92. Maslin, "Rival Divers," 11.

93. Discussions of the film in Susan Hayward and Phil Powrie, *The Films of Luc Besson: Master of Spectacle*, range from treating it as an example of Besson's neobaroque style, to postmodern pastiche, to Freudian allegory, to a study of melancholia, to mysticism, to, in Laurent Juillier's words, the description of "a self-destructive society, whose members knowingly decide to make fewer children than necessary to keep the population numbers at the same level" (Juillier, "The Sinking of the Self," 118).

94. Hayward, *Luc Besson*, 49.

95. Ibid., 11.

96. Juillier, "The Sinking of the Self," 111.

97. Mitman, *Reel Nature*, 169.

98. Quoted in Cimini, "'Blue Mind.'"

99. Sikov, *On Sunset Boulevard*, 299.

100. Ibid., 298.

101. Ibid.

102. Ibid., 299.

103. Ibid.

104. Beuka, "Just One . . . 'Plastics,'" 14.

105. Hamera, *Parlor Ponds*, 11.

106. Kashner, "Here's to You, Mr. Nichols."

107. Ibid.

108. Surtees, cited by Kashner, ibid.

109. Dobbs, "25 of Cinema's Most Memorable Swimming Pool Scenes."

110. D'Aloia, "Film in Depth," 97.

111. Ibid., 96.

112. Dargis, "Finding Magic."

113. Quoted in Ressner, "Steven Spielberg."

114. Scott, "Film Review: War Is Hell."

115. Rogers, "Pete Romano."

116. Quoted in Elrick, "Michael Bay's *Pearl Harbor.*"

117. Romano quoted in Rogers, "Pete Romano."

118. Quotations in this paragraph and the next are from Warshaw, *History of Surfing*, 291.

119. Pemberton, "Surfing in SLO."

120. *Big Wednesday*, Milius speaking in feature-length audio commentary, ca. 1:39:38.

121. Ibid.

122. The Poe poem was republished as "The City in the Sea" (1845), and Tourneur's film was named "The City under the Sea" for its UK release.

123. Blanchot, *Writing of the Disaster*, 58.

124. On Nemo's mute contemplation in *Twenty Thousand Leagues under the Seas* and its afterlife in recent astronaut films about climate catastrophe, see my "Helmeted Beholder."

125. Keller, *James Cameron*, 22.

126. Ibid.

127. Vivian Sobchack points to the "absolutely crucial *frame story*," but treats it primarily in figurative fashion. She emphasizes the metaphorical impact of the "*submersible* device that allows both viewers and characters in the present to slowly descend and relocate themselves in a 'lost' past history both public and private." Sobchack also notes that this deep-sea submersible is a "*spherical* device that encircles and contains both the past and present." Consistent with her interest in metaphors of enclosure, she sees the film's frame story as a "*protective* device," that "allows for a mediated but seemingly 'authentic' experience, one that keeps real trauma, but not real emotion, at bay." It remains, however, to integrate into these metaphorics Cameron's interest in showing the actual descent and the technologies used to capture such deep-sea imagery. Sobchack, "Bathos and the Bathysphere," 191–93.

128. Keller, *James Cameron*, 134.

129. Hardy, "New High-Powered Lighting."

130. Keegan, *Futurist*, 174.

131. Cameron, cited by Keegan, ibid.

132. Cameron, "Ghostwalking in Titanic," 100–101.

133. Cameron in the scene showing the NTIs also refers to the history of image technology and the use of film for surveillance, as Lindsey hangs up strips of analog film she has developed to look for glimpses of these computer-generated creatures.

134. Jessen, "Hydronicus Inverticus."

135. Henry Selick quoted in Vice, "Something Fishy," my emphasis.

136. Quoted in Jessen, "Hydronicus Inverticus."

137. Sean Hutchinson, "32 Facts about *The Life Aquatic*," 3.

138. All Selick quotations in this paragraph are from Vice, "Something Fishy."

139. This is the name painted on the side of the vessel. It is the name of Zissou's first wife, Jacqueline, which Zissou crosses out rather than effacing, in an ambivalent gesture suggesting at once his laziness and his attachment.

140. Selick discusses how hard it was to come up with the swimming motion for the animated puppets and mentions apropos of the jaguar shark that "our D.P., a guy named Pat Sweeney," had "shot a lot of those submarine movies, like *Hunt for Red October*," and knew how big models had to be "in order to move people through the scale" (quoted in Jessen, "Hydronicus Inverticus").

Epilogue

1. Bill Nichols usefully clarifies the range of documentary ethics and their history in *Introduction to Documentary*.

2. At some point in the 1950s, for example, Scientific Diving Consultants, based at Scripps Institution of Oceanography, advertised their availability to make such films in a brochure for the leading dive shop, The Diving Locker: http://scilib.ucsd.edu/sio/hist/diving_locker _scientific_diving_consultants_brochure.pdf (this link is no longer active, but the brochure is at the Scripps Institution of Oceanography Archive). Their film *"Rivers of Sand* [1959] . . . won many awards at film festivals throughout the world." Information from "James Stewart," at International Legends of Diving website. Scientific Diving Consultants' *Underwater Wonders* was also a crossover film.

3. This figure is from "Turning Trailblazing Science into Agenda-Setting TV."

4. Nichols, *Introduction*, 138.

5. Wilson, "Contact Zones," 719.

6. "Turning Trailblazing Science into Agenda-Setting TV."

7. Honeyborne, *"Blue Planet II*, An Introduction."

8. Halls, "Blue Planet 2: The Inside Story."

9. Beecham, "Blue Planet II: 'One Ocean.'"

10. Quoted in Halls, "Blue Planet 2: The Inside Story."

11. Reilly, "Blue Planet II Viewers Terrified."

12. Lambert, "Stunning Moment."

13. "Blue Planet II's Bird-Eating Fish."

14. Honeyborne, quoted in Christian, "Ingenious Technology."

15. Honeyborne, quoted in Nguyen, "'Planet Earth: Blue Planet II.'"

16. Honeybourne, quoted in Christian, "Ingenious Technology."

17. Hamilton, "10 of the Most Incredible Moments."

18. See Alaimo's "Violet-Black."

19. The credits for episode 2 thank James Cameron and Earthship Productions, a company Cameron formed after *Titanic*, in 1998, to pursue documentaries about marine environments.

20. See Elias, *Coral Empire*, 207.

21. As Stephen and Anthony Palumbi observe, *"Finding Nemo* gets many things right . . . but it punts away the most fascinating aspect of clownfish," which are all born male "with the ability to change sex." If Nemo's father lost his mate, Nemo's father would metamorphose into his mother, and "Nemo . . . would become his own father," enabling the pair to raise "little incestuous Nemos together without a drip of sentimentality" (*Extreme Life*, 141, 142).

22. Honeyborne, cited in Nguyen, "'Planet Earth: Blue Planet II.'"

23. Cormier, "Snapping Death Worms."

24. Lachat and Haag-Wackernagel, "Novel Mobbing Strategies."

25. Shepherd, "Blue Planet 2: Bobbit Worm Terrifies Viewers." Allen, "'Giant Carnivorous' Bobbit Worm."

26. Smith, "Filming Bobbit Worms."

27. Allen, "'Giant Carnivorous' Bobbit Worm."

28. Brownlow, cited in Nguyen, "'Planet Earth: Blue Planet II.'"

29. Wilson, "Contact Zones," 721.

30. Ibid., 720.

31. Richards, "Footage of a Mother Whale."

32. See Carrington, "Blue Planet 2." The BBC replied in "Blue Planet II: The Secrets."

33. Duell and Lambert, "BBC in New Fakery Row."

34. Honeyborne quoted in Carrington, "Blue Planet 2."

35. Ibid.

36. "Blue Planet II: The Secrets."

37. Honeyborne quoted in Carrington, "Blue Planet 2."

38. "Dive Deeper."

39. Doherty, "Filming Micro-Detail."

40. Ibid.

41. Bosiger, "Capturing the World's Bleaching Corals."

42. "Blue Planet II: The Secrets."

43. From Young, dir., *Making of 20,000 Leagues under the Sea*, Fleischer ca. 32:30.

44. "Blue Planet II: The Secrets."

45. Ibid.

46. Cronon, "Trouble with Wilderness," 81.

47. See, for example, Haraway's concept of "natureculture" introduced in *The Companion Species Manifesto*, Latour's concept of Gaia in *Facing Gaia*, and Morton, *Ecology without Nature*.

48. Wilson, "Contact Zones," 712ff.; Pratt, "Arts of the Contact Zone," 34.

49. Wilson, "Contact Zones," 720.

50. Royal Television Society, "Making of Blue Planet II," ca. 10:06.

51. Beecham, "Blue Planet II: 'One Ocean.'"

52. "Bird Eaters—Giant Trevally in the Seychelles."

53. Beecham, "Blue Planet II: 'One Ocean.'"

WORKS CITED

Abberley, Will, ed. *Underwater Worlds*. Cambridge: Cambridge Scholars Publishing, 2018.

Adamowsky, Natascha. *The Mysterious Science of the Sea, 1775–1943*. New York: Routledge, 2015.

Alaimo, Stacy. "Dispersing Disaster: The Deepwater Horizon, Ocean Conservation, and the Immateriality of Aliens." In *American Environments: Climate-Cultures-Catastrophe*, edited by Christof Mauch and Sylvia Mayer, 177–92. Heidelberg: University of Heidelberg Press, 2012.

———. "States of Suspension: Trans-corporeality at Sea." *Interdisciplinary Studies in Literature and Environment* 19, no. 3 (Summer 2012): 476–93.

———. "Violet-Black." In *Prismatic Ecology: Ecotheory beyond Green*, edited by Jeffrey Jerome Cohen, 233–51. Minneapolis: University of Minnesota Press, 2013.

Allen, Ben. "The 'Giant Carnivorous' Bobbit Worm Turns Blue Planet II into an Actual Horror Movie." *RadioTimes*, n.d. [November 13, 2017]. Accessed October 2020. https://www .radiotimes.com/news/tv/2018-02-09/the-giant-carnivorous-bobbit-worm-turns-blue -planet-ii-into-an-actual-horror-movie.

"All-Time Top Film Grossers [over 4,000,000 US-Canada Rentals]." *Variety*, January 8, 1964, 37.

Anthoni, J. Floor. "Underwater Photography." Seafriends. Accessed October 2020. http://www .seafriends.org.nz/phgraph.

Aron, Robert. "Films of Revolt." In *French Film Theory and Criticism 1907–1939*. Vol. 1, *1907–1929*, edited by Richard Abel, 432–36. Princeton, NJ: Princeton University Press, 1988.

"Author—R. M. Ballantyne." Classic Dive Books. Accessed October 2020. http://classicdive books.customer.netspace.net.au/oeclassics-a-ballantyne.html.

Ballantyne, R. M. *Under the Waves or Diving in Deep Waters—A Tale*. London: James Nisbet & Co., 1876.

Barrett, Eric. "Lotte Is Going Down Again." *Picturegoer*, October 24, 1953, 12.

Barry, Edward. "Film on Sponge Fishers Shows Beauty of Gulf." *Chicago Daily Tribune*, August 17, 1948, A4.

Barthes, Roland. *S/Z*. Translated by Richard Miller. Preface by Richard Howard. New York: Hill and Wang, 1974.

Bazin, André. "Le cinéma et l'exploration" (1958). In *Qu'est-ce que le cinéma?* 25–33. Paris: Editions du Cerf, 2000.

———. "Un film bergsonien: *Le mystère Picasso*" (1956). *Qu'est-ce que le cinéma?*, 193–202. Paris: Editions du Cerf, 2000.

———. "Information or Necrophagy" (1957). In *André Bazin's New Media* 124–25. Translated and edited by Dudley Andrew. Oakland: University of California Press, 2014.

———. "*Le monde du silence*" (1955). In *Qu'est-ce que le cinéma?*, 35–40. Paris: Editions du Cerf, 2000.

BBC Audience Research Department. "Report." April 24, 1956. Archive, Hans Hass Institut für Submarine Forschung und Tauchtechnik.

"BBC View." *Birmingham Mail*, May 12, 1956.

Bech, Janwilem. "Hans Hass, interview with Michael Jung." Accessed October 2020. http://www.therebreathersite.nl/Zuurstofrebreathers/German/hans_hass.htm.

Bedot, Maurice. *Hermann Fol, sa vie et ses travaux*. Geneva: Imprimerie Aubert-Schuchardt, 1894.

Beebe, William. *Beneath Tropic Seas: A Record of Diving among the Coral Reefs of Haiti*. New York: G. P. Putnam's Sons, 1928.

Beecham, Dan. "Blue Planet II: 'One Ocean'—Giant Trevallies Vs Terns." Accessed October 2020. http://www.danbeecham.com/afishoutofwater.

Béhar, Henri. *Le cinéma des surréalistes*. Cahiers du Centre de Recherche sur le Surréalisme Mélusine, no. 24. Lausanne: Editions de L'Age d'Homme, 2004.

Bellows, Andy Masaki, Marina McDougall, and Brigitte Berg, eds. *Science Is Fiction: The Films of Jean Painlevé*. Cambridge, MA: MIT Press, 2000.

"Beneath the Twelve-Mile Reef." *Variety*, December 16, 1954, 6.

Benton, Graham. "Shark Films: Cinematic Realism and the Production of Terror." In *This Watery World: Humans and the Sea*, edited by Nandita Batra and Vartan P. Messier, 123–32. Newcastle: Cambridge Scholars Publishing, 2008.

"The Best Scuba Diving Movies of All Time—Updated!" *Diver*, November 29, 2012. Accessed October 2020. http://divermag.com/the-best-diving-movies-of-all-time and the thread that follows.

Beuka, Robert. "Just One . . . 'Plastics': Suburban Malaise, Masculinity, and Oedipal Drive in *The Graduate*." *Journal of Popular Film and Television* 28, no. 1 (2000): 12–21.

Bie, Tom. "Bird Eaters—Giant Trevally in the Seychelles." *Drake*, December 13, 2017. Accessed October 2020. https://drakemag.com/bird-eaters.

"Biography: Pete Romano." HydroFlex. http://hydroflex.com/about-us/biography-pete-romano. Accessed October 2020.

Black, Peter. "Teleview." *Daily Mail*, April 28, 1956 (page number not on archive clipping).

Blanchon, Jean Henri. "Zarh Pritchard." Translation. *Exhibition of Undersea Paintings by Zarh Pritchard*, 3–4. Grace Nicholson Galleries. Pasadena, CA: Earl C. Tripp, 1926.

Blanchot, Maurice. *The Writing of the Disaster*. Translated by Ann Smock. Lincoln: University of Nebraska Press, 1995.

"Blue Planet II: The Secrets behind the BBC One Series." BBC News, October 25, 2017. https://www.bbc.com/news/entertainment-arts-41740841.

"Blue Planet II's Bird-Eating Fish Horrifies Viewers." *Telegraph*, October 30, 2017. https://www.telegraph.co.uk/tv/2017/10/30/blue-planet-iis-bird-eating-fish-horrifies-viewers.

Bordwell, David, Janet Staiger, and Kristin Thompson. *The Classical Hollywood Cinema: Film Style and Mode of Production to 1960*. New York: Columbia University Press, 1985.

Bosiger, Yoland. "Capturing the World's Bleaching Corals." BBC One. Accessed October 2020. https://www.bbc.co.uk/programmes/articles/1nfX6qBR3CZktzZJdjnmpdg/capturing-the-worlds-bleaching-corals.

Bousé, Derek. *Wildlife Films*. Philadelphia: University of Pennsylvania Press, 2000.

Boutan, Louis. "Emploi du scaphandre pour les études zoologiques et la photographie sous-marine." *Revue Scientifique* 1(1894):481–90.

———. *La photographie sous-marine et les progrès de la photographie*. Paris: Schleicher Frères, 1900.

Bowers, Q. David. "Volume II: Filmography. *The Terrors of the Deep*." Thanhouser Films: An Encyclopedia and History. 1995. Accessed October 2020. https://www.thanhouser.org /tcocd/Filmography_files/086kaj.htm.

Breton, André. *Mad Love*. Translated by Mary Ann Caws. Lincoln: University of Nebraska, Bison Books, 1988.

British Broadcasting Corporation. "An Audience Research Report." June 1, 1956. Archive, Hans Hass Institut für Submarine Forschung und Tauchtechnik.

Brunner, Bernd. *The Ocean at Home: An Illustrated History of the Aquarium*. Hudson, NY: Princeton Architectural Press, 2005.

Burgess, Thomas. *Take Me under the Sea: The Dream Merchants of the Deep*. Salem, MA: Ocean Archives, 1994.

Burke, Edmund. *A Philosophical Enquiry into the Origin of Our Ideas of the Sublime and the Beautiful* (1757), edited by Paul Guyer. New York: Oxford University Press, 2015.

Cahill, James. *Zoological Surrealism*. Minneapolis: University of Minnesota Press, 2019.

Calcagno, Claire. "Underwater Photography." Edgerton Digital Collections Project. Accessed October 2020. https://edgerton-digital-collections.org/stories/features/fathoming-the -oceans-2-under-water-photography.

Cameron, James. "Ghostwalking in Titanic." *National Geographic* 221, no. 4 (April 2012): 100–101. http://www.nationalgeographic.com/magazine/2012/04/titanic-shipwreck -underwater-exploration-james-cameron.

Canby, Vincent. "'The Deep,' a Movie, Is Shallow." *New York Times*, June 18, 1977, 10.

———. "Screen: Dramatic Pursuit of Elusive Killer Shark." *New York Times*, May 21, 1971, 51.

Carr, Winifred. "Glamour in a Strange Setting." No newspaper, date or page number in archive clipping. Carr wrote regularly for *The Daily Telegraph*, and the article references a show that aired on May 8, 1956.

Carrington, Damian. "Blue Planet 2: Attenborough Defends Shots Filmed in Studio." *Guardian*, October 23, 2017. https://www.theguardian.com/environment/2017/oct/23/captive -wildlife-footage-blue-planet-2-bbc1-totally-true-to-nature-say-producers.

Carroll, Noël. *The Philosophy of Horror: Or, Paradoxes of the Heart*. New York: Routledge, 1990.

Carter, Bill. "Media: TELEVISION; Stand Aside, CNN. America's No. 1 TV Export Is—No Scoffing, Please—'Baywatch.'" *New York Times*, July 3, 1995, 41.

Chancellor, Henry. *James Bond: The Man and His World*. Vicenza, Italy: Henry Chancellor and Ian Fleming Publications, 2005. First published in Great Britain in 2005 by John Murray, a division of Hodder Headline.

Chapman, David. "The Undersea World of the Sound Department: The Construction of Sonic Conventions in Sub-aqua Screen Environments." *New Soundtrack* 6, no. 2 (2016): 143–57.

Christian, Bonnie. "The Ingenious Technology behind Blue Planet II." *Wired*, November 26, 2017. https://www.wired.co.uk/article/blue-planet-2-technology-david-attenborough.

Christie, Ian. "Set Design." In *Encyclopedia of Early Cinema*, edited by Richard Abel. New York: Routledge, 2005.

Cimini, Marla. "'Blue Mind': Why Being near Water Makes You Happy." *USA Today,* November 13, 2017. https://www.usatoday.com/story/travel/destinations/2017/11/13/blue-mind/857903001.

Clarke, Gavin. "James Bond Doesn't Do CGI: Inside 007's Amazing Real-World Action." *Register,* October 25, 2012, https://www.theregister.co.uk/2012/10/25/bond_behind_the_scenes.

Cohen, Margaret. "Adventures in Toxic Atmosphere." In *Underwater Worlds,* edited by Will Abberley, 72–89. Cambridge: Cambridge Scholars Press, 2018.

———. "Denotation in Alien Environments: The Underwater *Je Ne Sais Quoi.*" *Representations* 125 (Winter 2014): 103–26.

———. "The Helmeted Beholder." *DIBUR,* issue 6, "Visions of the Future," Fall 2018. https:arcade.stanford.edu/dibur/helmeted-beholder.

———. *The Novel and the Sea.* Princeton, NJ: Princeton University Press, 2010.

———. "The Shipwreck as Undersea Gothic." In *The Aesthetics of the Undersea,* edited by Margaret Cohen and Killian Quigley, 155–66. New York: Routledge, 2019.

———. "The Underwater Imagination: From Environment to Film Set, 1954–1956." In "Hydrocriticism," edited by Laura Winkiel. Special issue, *English Language Notes* 57, no. 1 (2019): 51–71.

———. "Underwater Optics as Symbolic Form." *French Politics, Culture & Society* 32, no. 3 (Winter 2014): 1–23.

Cohen, Margaret, and Killian Quigley, eds. "Introduction." In *The Aesthetics of the Undersea,* 1–13. New York: Routledge, 2019.

Condee, William Faricy. *Theatrical Space: A Guide for Directors and Designers.* Lanham, MD: Scarecrow Press, 2001.

Conrad, Joseph. "Author's Note to *Typhoon*" (1919). In *Typhoon and Other Tales,* edited by Cedric Watts. New York: Oxford University Press, 2008.

"A Conversation with Zarh H. Pritchard, Recorded by the Editor." *Asia* (March 1924): 217–20.

Cooke, Mervyn. "Water Music: Scoring *The Silent World.*" In *Music and Sound in Documentary Film,* edited by Holly Rogers, 104–22. New York: Routledge, 2014.

Corbin, Alain. *The Lure of the Sea: The Discovery of the Seaside in the Western World 1750–1840.* Translated by Jocelyn Phelps. Berkeley: University of California Press, 1994.

Cormier, Zoe. "Snapping Death Worms Can Hide Undetected for Years." BBC Earth. Accessed September 2020. https://www.bbcearth.com/blog/?article=snapping-death-worms-can-hide-undetected-for-years.

Cousteau, Jacques-Yves. "Fish Men Explore a New World Undersea." *National Geographic* 102, no. 4 (October 1952): 431–70.

———. "Studio under the Sea: Sub-marine Scientist Scans Problems of Shooting in the Lower Depths." *New York Times,* September 16, 1956, X7.

———. *Window in the Sea.* Vol. 4 of *The Ocean World of Jacques Cousteau.* Rev. ed. Danbury, CT: Danbury Press, 1975.

Cousteau, Jacques-Yves, and Frédéric Dumas. *Le monde du silence* (1953). Paris: Editions de Paris, 1965.

Cousteau, Jacques-Yves, with Frédéric Dumas. *The Silent World: A Story of Undersea Discovery and Adventure, by the First Men to Swim at Record Depths with the Freedom of Fish.* New York: Harper & Row, 1953.

Cronon, William. "The Trouble with Wilderness; or, Getting Back to the Wrong Nature." In *Uncommon Ground: Rethinking the Human Place in Nature*, edited by William Cronon, 69–90. New York: W. W. Norton, 1995.

Crosby, Richard. "Cinematographer Dr. Hans Hass Takes Moviegoers 'Under the Red Sea.'" *Skin Diver*, January 1953, 8, cont. on 12.

Crowther, Bosley. "Screen: Beautiful Sea; 'Silent World' Opens at the Paris Here." *New York Times*, September 25, 1956, 30.

———. "Screen: 007's Underwater Adventures." *New York Times*, December 22, 1965, 23.

———. "The Screen in Review." [Review of *Reap the Wild Wind*.] *New York Times*, March 27, 1942, 27.

———. "The Screen in Review: 'Under the Red Sea' Reveals Wonders of the Piscatorial World at Beekman." *New York Times*, November 19, 1952, 37.

———. "The Screen: 'Underwater!'; Jane Russell Film Bows at Mayfair." *New York Times*, February 10, 1955, 27.

———. "The Screen: Wide-Angle Lens in Water: New Fox CinemaScope Movie, 'Beneath the 12-Mile Reef,' Presented at the Roxy." *New York Times*, December 17, 1953, 52.

Crylen, Jonathan. "The Cinematic Aquarium: A History of Undersea Film." PhD diss., University of Iowa, 2015.

Cubitt, Sean. *Ecomedia*. Amsterdam: Rodopi, 2005.

D'Aloia, Adriano. "Film in Depth: Water and Immersivity in the Contemporary Film Experience." *Acta Universitatis Sapientiae, Film and Media Studies* 5 (2102): 87–106.

"The Dangers of Underwater Swimming." *Yorkshire Post*, May 8, 1956, page not visible in archival clipping.

Dargis, Manohla. "Finding Magic Somewhere under the Pool." *New York Times*, July 21, 2006.

Dash, Thomas R. "The Frogmen." *Women's Wear Daily* (New York), July 2, 1951, 26.

Daugherty, Frank. "Arch Obeler Speculates Atomically." *Christian Science Monitor*, March 27, 1951, 11.

Davis, Robert H., Sir. *Deep Diving and Submarine Operations: A Manual for Deep Sea Divers and Compressed Air Workers, Parts I & II*. London: Saint Catherine Press, n.d. [1955].

Davis, Ronald L. *Duke: The Life and Image of John Wayne*. Norman: University of Oklahoma Press, 1998.

Deane, Robert. "Underwater Photography." In *Encyclopedia of Nineteenth-Century Photography*, 1416–17, edited by John Hannavy. New York: Routledge, 2008.

"The Deep." *Variety*, December 31, 1976. https://variety.com/1976/film/reviews/the-deep-2-1200423893.

"The Deep Reviews." *TV Guide*. Accessed October 2019. https://www.tvguide.com/movies/the-deep/review/112628.

DeLoughrey, Elizabeth. "Submarine Futures of the Anthropocene." *Comparative Literature* 69:1 (March 1, 2017): 32–44.

Dickens, Charles. "Night Walks." Originally "The Uncommercial Traveller." *All the Year Round* 3, no. 65 (July 21, 1860): 348–52, https://www.djo.org.uk/indexes/articles/the-uncommercial-traveller-no-12.html.

Dillow, Clay. "Video: The First Footage from James Cameron's Record-Setting Dive to Challenger Deep." *Popular Science*, March 27, 2012. https://www.popsci.com/technology/article/2012-03/video-first-footage-james-camerons-record-setting-dive-challenger-deep.

Dimmock, Kay, and Terry Cummins. "History of Scuba Diving Tourism." In *Scuba Diving Tourism*, edited by Ghazali Musa and Kay Dimmock, 14–28. New York: Routledge, 2013.

Diolé, Philippe. *The Undersea Adventure*. Translated by Alan Ross. London: Sidgwick & Jackson, 1953.

"Dive Deeper to See How Blue Planet II Was Made." BBC One. Accessed October 2020. https://www.bbc.co.uk/programmes/articles/cD5fGcsfj30hhFqJcfPQQR/dive-deeper-to -see-how-blue-planet-ii-was-made.

Dobbs, Sarah. "25 of Cinema's Most Memorable Swimming Pool Scenes." Den of Geek, November 2, 2015. Accessed October 2020. https://www.denofgeek.com/us/movies/old-school /250204/25-of-cinema-s-most-memorable-swimming-pool-scenes.

Doherty, Orla. "Filming Micro-Detail in the Deep." BBC One. Accessed October 2020. https:// www.bbc.co.uk/programmes/articles/2lrTgsZzqZnfQNMfZtFV8y1/filming-micro-detail -in-the-deep.

Dolbear, Sam. "Flooded Displays: The Arcades and the Aquarium" (first version). Paper given at "Working Worlds," UCLA Art History conference, May 2015. Accessed October 2020. https://www.academia.edu/20012816/Flooded_Displays_the_Arcade_and_the_Aquarium _Version_One.

Dowling, David. "'Revenge upon a Dumb Brute': Casting the Whale in Film Adaptations of *Moby-Dick*." *Journal of Film and Video* 66, no. 4 (Winter 2014): 50–63.

Duell, Mark, and Laura Lambert. "BBC in New Fakery Row as It Emerges Viewers WON'T Be Told When Scenes in Flagship Documentary Blue Planet II Were Filmed in Laboratories Rather Than the Wild." *Daily Mail*, October 23, 2017. https://www.dailymail.co.uk/news /article-5007471/Sir-David-Attenborough-91-says-NO-plans-retire.html.

Dyer, Sidney. *Ocean Gardens and Palaces, or, The Tent on the Beach*. Philadelphia: Benjamin Griffith, 1880.

Ecott, Tim. *Neutral Buoyancy: Adventures in a Liquid World*. New York: Grove Press, 2001.

Edge, Martin. *The Underwater Photographer*. 3rd ed. Oxford: Elsevier, 2008.

Eldredge, Charles C. "Wet Paint: Herman Melville, Elihu Vedder, and Artists Undersea." *American Art* 1, no. 2 (Summer 1997): 106–35.

Elias, Ann. *Coral Empire: Underwater Oceans, Colonial Tropics, Visual Modernity*. Durham, NC: Duke University Press, 2019.

———. "Frank Hurley and the Symbolic Underwater." In *The Aesthetics of the Undersea*, edited by Margaret Cohen and Killian Quigley, 124–36. New York: Routledge, 2019.

———. "Sea of Dreams: André Breton and the Great Barrier Reef." *Papers of Surrealism* 10 (Summer 2013): 1–15. https://www.research.manchester.ac.uk/portal/files/63517394 /surrealism_issue_10.pdf.

Ellis, Richard. *Shark: A Visual History*. Guilford, CT: Globe Pequot Press, 2012.

Elrick, Ted. "Michael Bay's Pearl Harbor." Michaelbay. Accessed April 2019. https://www .michaelbay.com/articles/michael-bays-pearlharbor/.

Erickson, Steven. "Animal Attraction." *Art Forum*, April 20, 2009, https://www.artforum.com /film/steven-erickson-on-jean-painleve-22565.

"Even the Secretary Goes under Water in Shooting 'Sea Hunt.'" *Variety*, February 5, 1958, 27, cont. on 50.

"Exception Form: November 25, 1965. *Thunderball*." British Board of Film Classification. Accessed April 2018. https://www.bbfc.co.uk/sites/default/files/attachments/thunderball

-final.pdf. These comments have subsequently been taken down. The BBFC offers the link press@bbfc.co.uk for research into documents concerning graphic sexual content.

Ferguson, Kathryn. "Submerged Realities: Shark Documentaries at Depth." *Atenea* 26, no. 1: 115–29.

Fernbach, Amanda. "The Fetishization of Masculinity in Science Fiction." *Science Fiction Studies* no. 81, vol. 27, pt. 2 (July 2000): https://www.depauw.edu/sfs/backissues/81/fernbach81art .htm.

Flipper trivia page. IMDb. Accessed October 2020. https://www.imdb.com/title/tt0057063 /trivia?ref_=tt_trv_trv.

Fol, Hermann. "Les impressions d'un scaphandrier." *La Revue Scientifique*, June 17, 1890. Golubik Sciences. Accessed October 2020. http://sciences.gloubik.info/spip.php?article1755.

Fox, Margalit. "Lotte Hass, Pioneering Diver Known for Beauty and Bravery, Dies at 86." *New York Times*, January 29, 2015, D7.

Frakes, Randall, Brooks Peck, Sidney Perkowitz, Matt Singer, Gary K. Wolfe, and Lisa Yaszek. *James Cameron's Story of Science Fiction.* San Rafael, CA: Insight Editions, 2018.

Fretz, Lauren E. "Surréalisme sous-l'eau: Science and Surrealism in the Early Films and Writings of Jean Painlevé." *Film & History* 40, no. 2 (Fall 2010): 45–65.

Fuchs, Derek. "Omnimax Film Attacks Shark Myths." TribLive, June 18, 2001. Accessed October 2018. https://triblive.com/x/pittsburghtrib/ae/museums/s_48259.html.

Gallant, Jeffrey. "First Underwater Photo." Diving Almanac. Accessed October 2020. http:// divingalmanac.com/first-underwater-photo/# (now inactive).

———. "John D. Craig," Diving Almanac. Accessed October 2020. http://divingalmanac.com /craig-john-d (now inactive).

Gevinson, Alan. *American Film Institute Catalog: Within Our Gates: Ethnicity in American Feature Films 1911–1960.* Berkeley: University of California Press, 1997.

Gilliam, Bret. "Al Giddings and the First Underwater Blockbuster Movie." *Fifty Fathoms: The Dive and Watch History 1953–2013,* edited by Dietmar Fuchs, 210–22. Switzerland: Watchprint.com for Blancpain, 2015.

———. "Al Giddings, Dean of Underwater Hollywood." SDI/TDI/ERDI/PFI. Accessed October 2020. https://www.tdisdi.com/diving-pioneers-and-innovators/al-giddings.

"Global Shark Conservation." Pew Charitable Trusts. Accessed October 2020. https://www .pewtrusts.org/en/projects/global-shark-conservation.

Gordon, Andrew M. *Empire of Dreams: The Science Fiction and Fantasy Films of Steven Spielberg.* Lanham, MD: Rowman & Littlefield, 2007.

Gosse, Philip Henry. *The Aquarium: An Unveiling of the Wonders of the Deep Sea* (1854). London: J. Van Vorst, 1856.

Griffin, Nancy, and Kim Masters. *Hit and Run: How Jon Peters and Peter Guber took Sony for a Ride in Hollywood.* New York: Touchstone, 1996.

Guber, Peter, with Barbara Witus. *Inside The Deep.* New York: Bantam, 1977.

Gunning, Tom. "The Cinema of Attraction[s]: Early Film, Its Spectator and the Avant-Garde." In *The Cinema of Attractions Reloaded,* edited by Wanda Strauven, 381–88. Amsterdam: Amsterdam University Press, 2006.

Haeckel, Ernst. "Professor Haeckel in Ceylon." *Nature* 26, no. 669 (August 24, 1882): 388–90.

———. *A Visit to Ceylon.* Translated by Clara Bell. 3rd American ed. New York: Peter Eckler, n.d. [1881?].

Halls, Eleanor. "Blue Planet 2: The Inside Story of the Bird-Eating Fish Scene." *GQ* online, October 29, 2017. https://www.gq-magazine.co.uk/article/blue-planet-2-fish-eating -birds.

Hamera, Judith. *Parlor Ponds: The Cultural Work of the American Home Aquarium, 1850–1970*. Ann Arbor: University of Michigan Press, 2012.

Hamilton, Isobel. "10 of the Most Incredible Moments from 'Blue Planet II.'" Mashable, December 15, 2017. Accessed September 2020. https://mashable.com/article/10-top-moments -blue-planet-ii.

Hanauer, Eric. *Diving Pioneers: An Oral History of Diving in America*. Locust Valley, NY: Aqua Quest Publications, 1999.

"Hans Hass." Wild Film History. Accessed April 2018. http://www.wildfilmhistory.org/person /102/Hans+Hass.html (now inactive).

Haraway, Donna. *The Companion Species Manifesto: Dogs, People, and Significant Otherness*. Chicago: Prickly Paradigm Press, 2003.

———. *When Species Meet*. Minneapolis: University of Minnesota Press, 2008.

Hardy, Kevin. "Return to the *Titanic*: The Third Manned Mission; New High-Powered Lighting, Dual Soviet Submarines, the Eerie Stillness of a Famous Wreck Promise to Bring Stunning Deep Ocean Images to IMAX Movie Audiences." *Sea Technology*, December 1991, n.p. https://www .deepsea.com/wp-content/uploads/ST_121991_Return_to_Titanic_Third_Manned_Mission .pdf.

Hass, Hans, and Michael Jung. *Aufbruch in eine neue Welt*. Königswinter: Heel Verlag, 2016.

Hayward, Susan. *Luc Besson*. Manchester: Manchester University Press, 1998.

Heberlein, Hermann. "Historical Development of Diving and Its Contribution to Marine Science and Research." *Proceedings of the Royal Society of Edinburgh, Section B. Biology* 72, no. 1 (1972): 283–96. Published online December 5, 2011, by Cambrige University Press: https://www.cambridge.org/core/journals/proceedings-of-the-royal-society-of-edinburgh -section-b-biological-sciences/article/abs/27historical-development-of-diving-and-its -contribution-to-marine-science-and-research/FB531118081F11F8CF2DAA402BB36D9A.

"Here's Looking at You, 1936–1939." History of the BBC, part 3. Teletronic: The Television History Site. Accessed October 2020. http://www.teletronic.co.uk/herestv2.htm.

Honeyborne, James. "*Blue Planet II, An Introduction*." BBC, October 15, 2017. https://www.bbc .co.uk/mediacentre/mediapacks/blue-planet-ii.

Hutchinson, Sean. "32 Facts about *The Life Aquatic with Steve Zissou*." Mental Floss, December 24, 2014. http://mentalfloss.com/article/60795/32-facts-about-life-aquatic-steve-zissou.

———. "25 Incisive Facts about *Jaws*." Mental Floss, June 20, 2018. http://mentalfloss.com /article/64548/25-incisive-facts-about-jaws.

Hutchinson, Tom. "20,000 Headaches." *Picturegoer*, July 30, 1955, 12–13.

Ilardo, Melissa, et al. "Physiological and Genetic Adaptations to Diving in Sea Nomads." *Cell* 173, no. 3 (April 19, 2018): 569–80.

Jackson, Matthew. "10 Fascinating Facts about *Thunderball*." Mental Floss, December 23, 2015. Accessed October 2020. https://www.mentalfloss.com/article/72999/10-fascinating-facts -about-thunderball.

"James Stewart." International Legends of Diving. Accessed September 2020. http://www .internationallegendsofdiving.com/FeaturedLegends/Jim_Stewart_bio.htm (now inactive).

Jay, Martin. *Downcast Eyes: The Denigration of Vision in Twentieth-Century French Thought.* Berkeley: University of California Press, 1993.

Jessen, Taylor. "Hydronicus Inverticus: An Interview with Henry Selick." ANIMATIONWorld, December 15, 2004. https://www.awn.com/animationworld/hydronicus-inverticus-interview-henry-selick.

Johnson, Martin W., F. Alton Everset, and Robert W. Young. "The Role of Snapping Shrimp (*Crangon* and *Synalpheus*) in the Production of Underwater Noise in the Sea." *Biological Bulletin* 93, no. 2 (October 1947): 122–138.

Johnson, Paul. "Sex, Snobbery, and Sadism." *New Statesman*, April 5, 1958. https://www.newstatesman.com/society/2007/02/1958-bond-fleming-girl-sex.

Jovanovic-Kruspel, Stefanie, Valérie Pisani, and Andreas Hantschk. "'Under Water'—Between Science and Art—The Rediscovery of the First Authentic Underwater Sketches by Eugen von Ransonnet-Villez (1838–1926)." *Annalen des Naturhistorischen Museums im Wien, Serie A* 119 (February 15, 2017): 131–53.

Juillier, Laurent. "The Sinking of the Self: Freudian Hydraulic Patterns in *Le Grand Bleu*." In *The Films of Luc Besson: Master of Spectacle*, edited by Susan Hayward and Phil Powrie, 109–20. New York: Manchester University Press, 2006.

"Juvenile Books—Novels, Fiction, Fact and Other Underwater Stories." Classic Dive Books. Accessed October 2020. http://classicdivebooks.customer.netspace.net.au/oeclassics-juvenile.html.

Kashner, Sam. "Here's to You, Mr. Nichols: The Making of *The Graduate*." *Vanity Fair*, March 2008. https://archive.vanityfair.com/article/2017/11/heres-to-you-mr-nichols-the-making-of-the-gradaute (now inactive).

Keegan, Rachel. *The Futurist: The Life and Films of James Cameron.* New York: Three Rivers Press, 2010.

Keller, Alexandra. *James Cameron.* New York: Routledge, 2006.

Kendall, Henry, and Hilbert Schenck. *Underwater Photography.* Cambridge, MA: Cornell Maritime Press, 1954.

Kenigsberg, Ben. "Tom Cruises's Most Dangerous Stunts in *Mission: Impossible*." *New York Times*, July 30, 2018. https://www.nytimes.com/2018/07/30/movies/mission-impossible-fallout-stunts-tom-cruise.html.

Kirkham, Pat. "Dots and Sickles: Maurice Binder's Film Titles." *Sight and Sound*, December 1995, 10–12.

Krauss, Rosalind. "Corpus Delicti." In *L'Amour fou: Photography and Surrealism*, edited by Rosalind Krauss, Dawn Ades, and Jane Livingston, 55–112. New York: Abbeville Press, 1985.

Kroll, Gary. *America's Ocean Wilderness: A Cultural History of Twentieth-Century Exploration.* Lawrence: University Press of Kansas, 2008.

Lachat, Jose, and Daniel Haag-Wackernagel. "Novel Mobbing Strategies of a Fish Population against a Sessile Annelid Predator." *Scientific Reports* 6, 33187 (2016).

Lamb, Jonathan. *Preserving the Self in the South Seas: 1680–1840.* Chicago: University of Chicago Press, 2001.

Lambert, Laura. "Stunning Moment Giant Trevally Fish Breaks from the Water and Plucks a Bird from the Sky." *Daily Mail*, October 22, 2017. https://www.dailymail.co.uk/news/article-5006947/Giant-trevally-fish-eats-bird-sky-Blue-Planet-II.html.

Latour, Bruno. *Facing Gaia: Eight Lectures on the New Climatic Regime*. New York: Polity, 2017.

Leonardo da Vinci. *Notebooks*. Selected by Irma A. Richter, edited and introduced by Thereza Wells, preface by Martin Kemp. New York: Oxford University Press, 2008.

"*The Life Aquatic with Steve Zissou*," IMDb. Accessed April 2019. https://www.imdb.com/title/tt0362270.

"Little-Known Navy Heroes Pictured in 'The Frogmen': They're Underwater Demolition Teams." *New York Herald Tribune*, June 24, 1951, D4.

"Lloyd Bridges Aqualung Actor." *TV Guide*, September 27, 1958, 21–23.

"Lloyd Bridges' Big Splash in *Sea Hunt*." *TV Guide*, June 27–July 3, 1959, 17–19.

Luria, S. M., and Jo Ann S. Kinney. "Underwater Vision." *Science* 167 (1970): 1454–61.

Magrini, James M. "'Surrealism' and the Omnipotence of Cinema." *Senses of Cinema* 44 (August 2007). Accessed October 2020. http://sensesofcinema.com/2007/feature-articles/surrealism-cinema.

Malle, Louis. *Malle on Malle*, edited by Philip French. London: Faber & Faber, 1993.

Mallinckrodt, Rebekka von. "Exploring Underwater Worlds: Diving in the Late Seventeenth-/Early Eighteenth-Century British Empire." In *Empire of the Senses: Sensory Practices of Colonialism in Early America*, edited by Daniela Hacke and Paul Musselwhite, 300–322. Boston: Brill, 2017.

Marden, Luis. "Camera under the Sea." *National Geographic*, February 1956, 162–200.

Marx, Robert F. *The History of Underwater Exploration* (1978). New York: Dover Publications, 1990.

Maslin, Janet. "Rival Divers Brave the Depths of the Sea." *New York Times*, August 20, 1988, 11.

McDonnell, Brian. "Film Noir Style." In *Encyclopedia of Film Noir*, edited by Geoff Mayer and Brian McDonnell, 70–81. Westport, CT: Greenwood Publishing Group, 2007.

Mentz, Steve. *At the Bottom of Shakespeare's Ocean*. London: Continuum, 2009.

Mercado, Gustavo. *The Filmmaker's Eye*. Burlington: MA: Focal Press (Elsevier), 2011.

Milman, Oliver. "The Deep Ocean: Plunging to New Depths to Discover the Largest Migration on Earth." *Guardian*, August 17, 2016. https://www.theguardian.com/environment/2016/aug/17/ocean-research-marine-life-bermuda-coral-reefs-nekton-triton-vessel.

Milne-Edwards, Henri. "Rapport adressé à M. le ministre de l'instruction publique, par M. Milne-Edwards, Membre de l'Institut, chargé d'une mission scientifique en Sicile." Paris: Imprimerie Paul Dupont, 1844.

Mitman, Gregg. *Reel Nature: America's Romance with Wildlife on Film*. Seattle: University of Washington Press, 2009.

Molesworth, Jesse. "Gothic Time, Sacred Time." *Modern Language Quarterly* 75, no. 1 (2014): 29–55.

Moody, Paul. *EMI Films and the Limits of British Cinema*. New York: Palgrave MacMillan, 2018.

Morris, Alan. "Sea-Bed Thrills for Viewers." *Press and Journal* (Aberdeen), April 3, 1956, page not visible in archival clipping.

———. "TV Remembers the Divers." *Sheffield Telegraph*, April 4, 1956, 3.

Morton, Timothy. *Ecology without Nature: Rethinking Environmental Aesthetics*. Cambridge MA: Harvard University Press, 2009.

Moure, Nancy Dustin Wall. *The World of Zarh Pritchard*. Carmel, CA: William A. Karges Fine Art, 1999.

Mulvey, Laura. "Visual Pleasure and Narrative Cinema." In *Film Theory and Criticism: Introductory Readings*, edited by Leo Braudy and Marshall Cohen, 833–44. New York: Oxford University Press, 1999.

"New Film, Modern Theatre, *16 Fathoms Deep.*" *Daily Boston Globe*, August 12, 1948, 10.

Newman, John Henry. "The Dream of Gerontius" (1865). Newman Reader—Works of John Henry Newman. Accessed October 2020. http://www.newmanreader.org/works/verses /gerontius.html.

Nguyen, Hanh. "'Planet Earth: Blue Planet II': 8 Ways Producers Filmed the Wet and Wild Aquatic Stars." *IndieWire*, January 18, 2018. https://www.indiewire.com/2018/01/planet-earth -blue-planet-2-cameras-amc-bbc-america-1201918989/.

Nichols, Bill. *Introduction to Documentary*. Bloomington: University of Indiana Press, 2001.

"1956 Le monde du silence." Cannes Film Festival. Accessed October 2020. http://www.cannes -fest.com/1956.htm.

North, Max. "The Battle of the Beaches Is ON." *Manchester Evenings News*, April 6, 1956, page not visible in archival clipping.

Norton, Trevor. *Underwater to Get out of the Rain: A Love Affair with the Sea*. Boston: Da Capo Press, 2007.

Oetterman, Stephan. Translated by Deborah Lucas Schneider. *The Panorama: History of a Mass Medium*. New York: Zone Books, 1997.

Official Guide New York World's Fair, 1965/1965. Time-Life Books, 1964.

O'Hanlan, Sean. "The Shipwreck of Reason: The Surrealist Diver and Modern Maritime Salvage." In *The Aesthetics of the Undersea*, edited by Margaret Cohen and Killian Quigley, 137–54. New York: Routledge, 2019.

"On Location with *Jaws.*" *American Cinematographer*, June 3, 2020. https://ascmag.com/articles /on-location-with-jaws.

Painlevé, Jean. "The Seahorse." In *Science Is Fiction: The Films of Jean Painlevé*, edited by Andy Masaki Bellows, Marina McDougall, and Brigitte Berg, xiii. Cambridge, MA: MIT Press, 2001.

Palumbi, Stephen R., and Anthony R. Palumbi. *The Extreme Life of the Sea*. Princeton, NJ: Princeton University Press, 2014.

Panofsky, Erwin. *Perspective as Symbolic Form*. Translated by Christopher S. Wood. New York: Zone Books, 1997.

Parisi, Paula. "Lunch on the Deck of the Titanic." *Wired*, February 1, 1998. https://www.wired .com/1998/02/cameron-3.

Parry, Zale, and Albert Tillman. *Scuba America: The Human History of Sport Diving*. Olga, WA: Whalestooth Publishing, 2011. Kindle.

Pemberton, Patrick S. "Surfing in SLO: Gary Busey, Others Talk Surf Movie 'Big Wednesday.'" *Tribune*, March 5, 2015. https://www.sanluisobispo.com/entertainment/movies-news -reviews/article39514887.html.

Pepys, Tom. "Underwater Grace, but a Clumsy Script." *Widnes Weekly News*, May 4, 1956, no page number in archive clipping.

Pitts, Michael R. *Columbia Pictures: Horror, Science Fiction and Fantasy Films, 1928–1982*. Jefferson, NC: McFarland, 2010.

Pound, Reginald. "Critic on the Hearth." *Listener*, April 12, 1956, 424.

Powell, Helen. *Stop the Clocks!: Time and Narrative in Cinema*. New York: I. B. Tauris, 2012.

Pratt, Mary Louise. "Arts of the Contact Zone." *Profession* (1991): 33–40.

Price, Mary Lynn. "A Biography of Conrad Limbaugh." ScubaBoard, May 2, 2008. Accessed April 2020. https://www.scubaboard.com/community/threads/connie-limbaugh-scientist -scuba-training-pioneer.561397.

Quick, Dan. "A History of Closed Circuit Oxygen Underwater Breathing Apparatus." Project 1/70, School of Underwater Medicine, HMAS *Penguin*, Balmoral, Australia, May 1970. https://allthingsdiving.com/download/a-history-of-closed-circuit-oxygen-underwater -breathing-apparatus-ccuba-dan-quick-1970.

Quigley, Killian. "The Porcellanous Ocean: Matter and Meaning in the Rococo Undersea." In *The Aesthetics of the Undersea*, edited by Margaret Cohen and Killian Quigley, 28–41. New York: Routledge, 2019.

Radcliffe, Ann. *The Mysteries of Udolpho* (1794). New York: Penguin, 2001.

"The Raft," *Sea Hunt*, episode identification 72B, April 21, 1959. Script by Art Arthur. In T. Robert Kendall Papers, Scripps Institution of Oceanography, La Jolla, CA.

Raimondo-Souto, H. Mario. *Motion Picture Photography: A History, 1891–1960.* Jefferson, NC: McFarland, 2007.

Raines, Halsey. "Movie Log of a Famed Fish Story: Hemingway's 'Old Man' Sails a Rough Route from Book to Film." *New York Times*, October 5, 1958, 7.

Ransonnet-Villez, Eugen, Baron von. *Ceylon, Skizzen seiner Bewohner, seines Thier- und Pflanzenlebens and Untersuchungen des Meeresgrundes nahe der Küste.* Braunschweig: George Westermann, 1868.

———. *Reise von Kairo nach Tor zu den Korallenbänken des rothen Meeres.* Vienna: Carl Ueberreuter, 1863.

———. *Sketches of the Inhabitants, Animal Life and Vegetation in the Lowlands and High Mountains of Ceylon, as Well as of the Submarine Scenery near the Coast Taken in a Diving Bell.* Vienna: Printed for the author by Gerold & sold by Robert Hardwicke, London, 1867.

"*Reap the Wild Wind*: Notes." Turner Classic Movies. Accessed October 2020. https://www.tcm .com/tcmdb/title/4375/reap-the-wild-wind#notes.

"*Reap the Wild Wind*: Production Notes." Bonus material with Universal DVD.

Reilly, Nick. "Blue Planet II Viewers Terrified by Bird Eating Fish." NME.com, October 30, 2017. https://www.nme.com/news/blue-planet-ii-viewers-amazed-by-bird-eating-fish-2154552.

Ressner, Jeffrey. "Steven Spielberg: War Is Hell." *DGA Quarterly* (Fall 2011). https://www.dga .org/Craft/DGAQ/All-Articles/1103-Fall-2011/Shot-to-Remember-Saving-Private-Ryan .aspx.

Rich, Adrienne. "Diving into the Wreck." In *Diving into the Wreck: Poems 1971–1972.* New York: W.W. Norton, 1973. Accessed October 2020. https://www.poets.org/poetsorg/poem/diving -wreck.

Richards, Alexandra. "Footage of a Mother Whale Carrying Her Dead Calf Has Made Blue Planet II Viewers Vow Never to Use Plastic Again." *Evening Standard*, November 20, 2017. Republished at BusinessInsider.com. Accessed October 2020. https://www.businessinsider .com/dead-whale-calf-shocks-blue-planet-ii-viewers-2017-11.

Rodríguez-Rincón, Luis. "The Aesthetics of the Early Modern Grotto and the Advent of an Empirical Nature." In *The Aesthetics of the Undersea*, edited by Margaret Cohen and Killian Quigley, 14–27. New York: Routledge, 2019.

Rogers, Pauline. "Pete Romano—Underwater Shooting for Pearl Harbor." *ICG Magazine*, May 2001. http://hydroflex.com/pete-romano-underwater-shooting-for-pearl-harbor/.

"Rolleimarin Underwater Camera Owned by Neville Collins Photographer." Accessed April 2018. http://harrington.server101.com/proddetail.php?prod=rolleimarin (now inactive).

Ross, Helen E. *Behaviour and Perception in Strange Environments*. London: George Allen & Unwin, 1974.

———. "Mist, Murk, and Visual Perception." *New Scientist* 66, no. 954 (1975): 658–60.

Rozwadowski, Helen M. "Arthur C. Clarke and the Limitations of the Ocean as a Frontier." *Environmental History* 17, no. 3 (2012): 578–602.

———. *Fathoming the Ocean: The Discovery and Exploration of the Deep Sea*. Paperback ed. Cambridge, MA: Belknap Press of Harvard University Press, 2008.

———. "Ocean's Depths." *Drunken Boat* 16. Accessed October 2019. https://d7drunkenboat .com/db16/helen-rozwadowski.html.

———. "Playing by—and on and under—the Sea: The Importance of Play for Knowing the Ocean." In *Knowing Global Environments: New Historical Perspectives on the Field Sciences*, edited by Jeremy Vetter, 162–89. New Brunswick, NJ: Rutgers University Press, 2010.

Rushing, Janice Hocker, and Thomas S. Frentz. *Projecting the Shadow: The Cyborg Hero in American Film*. Chicago: University of Chicago Press, 1995.

Schallert, Edwin. "'16 Fathoms Deep' Has Picturesque Appeal." *Los Angeles Times*, August 5, 1948, 21.

Schiller, Friedrich. "Der Taucher" [The Diver] (1797). Translated by Edward Bulwer-Lytton. Accessed November 2020. http://www.bartleby.com/360/9/10.html.

Scott, A. O. "Film Review; War Is Hell, but Very Pretty." *New York Times*, May 25, 2001.

"Sea Hunt." IMPDb. http://www.impdb.org/index.php?title=Sea_Hunt. Accessed April 2018.

Shakespeare, William. *The Tempest* (1623). Accessed October 2020. Folger Shakespeare Library. https://shakespeare.folger.edu/shakespeares-works/the-tempest.

"Shark Week 2019 Announced: A Summer Event of Bigger Sharks and Bigger Bites." Discovery Channel. Accessed November 2019. https://www.discovery.com/shark-week/shark-week -annoucnement-2019 (now inactive).

Shepard, Richard F. "Flipper, the Educated Dolphin, Cavorts in a Seascape Drama." *New York Times*, September 19, 1963, 23.

Shepherd, Jack. "*Blue Planet* 2: Bobbit Worm Terrifies Viewers of BBC One Show." *Independent*, November 13, 2017. https://www.independent.co.uk/arts-entertainment/tv/news/blue -planet-2-bobbit-worm-twitter-viewers-a8051796.html.

Shick, J. Malcolm. "Otherworldly." In *Underwater*, edited by Sanna Moore. Catalog for a Towner Touring exhibition curated by Angela Kingston, 33–39. Eastbourne: Towner Art Gallery, 2010.

Shor, Elizabeth. "Zarh H. Pritchard." UC San Diego. Accessed October 2020. http://scilib.ucsd .edu/sio/hist/Shor-Pritchard.pdf.

Sidney, George. "Diving with 'Jupiter's Darling.'" *New York Times*, November 14, 1954, 5.

Sikov, Ed. *On Sunset Boulevard: The Life and Times of Billy Wilder*. Reprint, Jackson: University Press of Mississippi, 2017. First published 1998 by Hyperion Press.

Sitney, P. Adams. *Modernist Montage: The Obscurity of Vision in Cinema and Literature*. New York: Columbia University Press, 1990.

"16 Fathoms Deep." *Times of India*, November 18, 1949, 9.

"16 Fathoms Deep." *Variety*, June 9, 1948, 12.

"Sixteen Fathoms Deep." *Variety*, January 23, 1934, 113.

Slide, Anthony. *The Encyclopedia of Vaudeville*. Reprint, Jackson: University Press of Mississippi, 2012. First published 1994 by Greenwood Press.

Smith, Jonathan. "Filming Bobbit Worms in the Dark." Accessed October 2020. https://www.bbc.co.uk/programmes/articles/1zzBxvhrqQRR4gpj7YG5ZjW/filming-bobbit-worms-in-the-dark.

Sobchack, Vivian. "Bathos and the Bathysphere: On Submersion, Longing, and History." In *"Titanic": Anatomy of a Blockbuster*, edited by Kevin Sandler and Gaylyn Studlar, 189–204. New Brunswick, NJ: Rutgers University Press, 1999.

Sorensen, Chris. "Ottawa Shows Off Bell Recovered from HMS Erebus." *Maclean's*. Accessed October 2020. http://www.macleans.ca/news/canada/ottawa-shows-off-bell-recovered-from-hms-erebus/.

Stafford, Jeff. "Reap the Wild Wind." October 6, 2006. Turner Classic Movies. Accessed October 2020. https://www.tcm.com/tcmdb/title/4375/reap-the-wild-wind#articles-reviews?articleId=147262.

Starosielski, Nicole. "Beyond Fluidity: A Cultural History of Cinema under Water." In *Ecocinema Theory and Practice*, edited by Salma Monani, Stephen Rust, and Sean Cubitt, 149–68. New York: Routledge, 2012.

Stevenson, Robert Louis. "Random Memories: The Education of an Engineer." In *The Works of Robert Louis Stevenson* (London: Chatto and Windus, 1912), vol. 16, 167–76. https://www.gutenberg.org/files/30990/30990-h/30990-h.htm.

Stott, Rebecca. *Darwin and the Barnacle: The Story of One Tiny Creature and History's Most Spectacular Scientific Breakthrough*. New York: W.W. Norton, 2004.

Straten, Frank Van. "Annette Kellerman 1886–1975." Live Performance Australia Hall of Fame, 2007. Accessed October 2018. https://liveperformance.com.au/halloffame/annettekellerman2.html (now inactive).

"The Sublime, the Picturesque, the Beautiful." For *American Scenery: Different Views in Hudson River School Painting*. Austin, TX: Blanton Museum of Art, 2012. Exhibition catalog. Accessed December 2018. https://blantonmuseum.org/files/american_scenery/sublime_guide.pdf (now inactive).

"Submarine Photography." *Photographic Journal of America* 58, no. 12 (December 1921): 477–80.

Sweeney, Kenneth. "Jaws (1975)." *American Cinematographer*, October 2012. https://theasc.com/ac_magazine/October2012/DVDPlayback/page1.html.

Sylph, Ann. "The Fish House at the ZSL London Zoo—the First Public Aquarium." Zoological Society of London, May 30, 2018. https://www.zsl.org/blogs/artefact-of-the-month/the-fish-house-at-zsl-london-zoo-the-first-public-aquarium.

Syme, Alison. *A Touch of Blossom: John Singer Sargent and the Queer Flora of Fin-de-Siècle Art*. University Park: Pennsylvania State University Press, 2010.

Taves, Brian. "A Pioneer under the Sea, Library Restores Rare Footage." *Information Bulletin* 55, no. 15, Library of Congress, September 16, 1996. https://www.loc.gov/loc/lcib/9615/sea.html.

The Thistlegorm Project. Accessed October 2020. https://thethistlegormproject.com.

"Topics." *New York Times*, July 18, 1960, 26.

Torma, Franziska. "Frontiers of Visibility: On Diving Mobility in Underwater Films (1920s to 1970s)." *Transfers* 3, no. 2 (June 2013): 24–46.

"Tors Envisions $1,000,000 Take from Sea Hunt." *Variety*, April 1, 1959, 30.

"Turning Trailblazing Science into Agenda-Setting TV." UKRI National Environmental Research Council. Accessed September 2020. https://nerc.ukri.org/research/impact/casestudies/society/tv.

"*20,000 Leagues under the Sea*." AFI Catalogue of Feature Films. Accessed November 2020. https://catalog.afi.com/Catalog/moviedetails/51389.

"Under the Red Sea." *Variety*, October 1, 1952, 22.

"Unsung Heroes of Naval Action." *New York Times*, June 30, 1951, 8.

[Untitled on *Diving to Adventure*]. *Birmingham Mail*, April 21, 1956, no page number in archive clipping.

Vaughan, Don. "Chesley Bonestell: Imagining the Future." *Print*, July 20, 2018. https://www.printmag.com/uncategorized/chesley-bonestell-imagining-the-future.

Vaughan, Malcolm. "Painting Beauty under the Sea." *Los Angeles Times*, July 8, 1928, 16–17.

Verne, Jules. *Twenty Thousand Leagues under the Seas*. Translated by William Butcher. Oxford: Oxford University Press, 1998.

Vice, Jeff. "Something Fishy—'The Life Aquatic' Features 'Animated' Marine Life." *Deseret News*, December 24, 2004. https://www.deseret.com/2004/12/24/19868196/something-fishy-151-the-life-aquatic-features-animated-marine-life#stop-motion-animator-henry-selick.

Vigo, Jean. "Towards a Social Cinema." Translated by Stuart Liebman. *Sabzian*, June 22, 2015. https://www.sabzian.be/article/toward-a-social-cinema.

"Virtual Tour of Second World War Ship Goes Online." *Guardian*, October 6, 2017. https://www.theguardian.com/world/2017/oct/06/virtual-tour-of-second-world-war-shipwreck-Thistlegorm-goes-online.

Vitello, Paul. "Hans Hass, Early Explorer of the World Beneath the Sea." *New York Times*, July 7, 2013, A18.

Voger, Mark. "Ricou Browning Performed Swimmingly as Movie Monster." *New Jersey Advance Media* for NJ.com, April 26, 2013. http://www.nj.com/entertainment/2013/04/creature_from_the_black_lagoon.html.

Walcott, Derek. "The Schooner *Flight*" (1979). In *Collected Poems 1948–1984*. New York: Farrar, Straus & Giroux, 1986. Accessed October 2020. https://www.poetryfoundation.org/poems/48316/the-schooner-flight.

Wall-Romana, Christophe. *Cinepoetry: Imaginary Cinemas in French Poetry*. New York: Fordham University Press, 2012.

Warshaw, Matt. *The History of Surfing*. San Francisco: Chronicle Books, 2010.

Watt, John. "Deep Sea 'Ballet' Made Good Viewing." *Evening Dispatch* (Edinburgh), April 28, 1956, no page number in archive clipping.

Watts, Stephen. "Activities on Britain's Varied Film Fronts." *New York Times*, December 1, 1957, X5.

"Weeki Wachee's History." Weeki Fresh Water Adventures. Accessed October 2020. https://weekiwachee.com/about-us/history/.

"A Weird New Film World." *Life* 36, no. 8 (February 22, 1954): 111–17.

Weiss, Sylvia. "Killers of the Sea." *Billboard*, July 24, 1937, 23.

Wells, H. G. "In the Abyss" (1896). In *The Plattner Story and Others*, 71–93. London: Methuen, 1897.

Wild, Jennifer. "For a Concrete Aesthetics: Against Avant-Grade Film c. 1930." *Framework* 58, nos. 1 and 2 (Spring/Fall 2017): 67–78.

Wilkinson, Leo. "Inside James Bond's Lotus Supersub." *Telegraph*, August 12, 2013. https://www.telegraph.co.uk/motoring/news/10233869/Inside-James-Bonds-Lotus-supersub.html.

Williams, Rosalind. *The Triumph of Human Empire: Verne, Morris, and Stevenson at the End of the World*. Chicago: University of Chicago Press, 2013.

Willson, Robert. "*Jaws* as Submarine Movie." *Jump Cut* 15 (1977): 32–33.

Wilson, Helen F. "Contact Zones: Multispecies Scholarship through *Imperial Eyes*." *Environment and Planning E: Nature and Space* 2, no. 4 (2019): 712–31.

Williamson, J. E. *Twenty Years Under the Sea*. New York: Hale, Cushman & Flint, 1936.

Winkiel, Laura. "Hydrocriticism." *English Language Notes* 57, no. 1 (2019): 1–10.

"Wreck Diving: Wrecks of the Red Sea; Thistlegorm," Chamber of Diving and Water Sports (Egypt). Accessed October 2020. https://www.cdws.travel/wreck-diving/thistlegorm.

"Wreck of the 'Télémaque.'" *Mechanics Magazine, Museum, Register Journal and Gazette*, July 2–December 31, 1842, vol. 37, pp. 186–87. London: J. C. Robertson, n.d.

Wright, Chris. "James Bond Retrospective: Thunderball (1965)." WhatCulture, July 26, 2012. http://whatculture.com/film/james-bond-retrospective-thunderball-1965.

Wulff, Hans J. "Suspense and the Influence of Cataphora on Viewers' Expectations." derwulff.de. Accessed October 2020. http://www.derwulff.de/files/2-58.pdf. A first version of this article appeared in *Suspense: Conceptualizations, Theoretical Analyses, and Empirical Explorations*, edited by Peter Vorderer, Hans J. Wulff, and Mike Friedrichsen, 1–17. Hillsdale, NJ: Lawrence Erlbaum, 1996.

Yamashiro, Shin. *American Sea Literature: Seascapes, Beach Narratives, and Underwater Exploration*. New York: Palgrave MacMillan, 2014.

Films and TV Shows Cited

Allen, Irving, dir. *16 Fathoms Deep*. 1948. Los Angeles: Irving Allen Productions, distributed by Monogram Pictures. Zeus. DVD.

Anderson, Wes, dir. *The Life Aquatic with Steve Zissou*. 2004. New York: Criterion Collection, 2005. DVD.

Arnold, Jack, dir. *Creature from the Black Lagoon*. 1954. Universal City, CA: Universal Pictures, 2000. DVD.

Bacon, Lloyd, dir. *The Frogmen*. 1951. Los Angeles: 20th Century Studios, 2005. DVD.

Bay, Michael, dir. *Pearl Harbor*. 2001. Burbank, CA: Touchstone Pictures, 2001. DVD.

Besson, Luc, dir. *The Big Blue [Le grand bleu]*. 1988. UK: Optimum World, 2009. DVD.

Bigelow, Kathryn, dir. *Point Break*. 1991. Burbank, CA: Warner Bros., 2017. DVD.

Bruno, John, Ray Quint, and Andrew Wight, dirs. *Deepsea Challenge 3D*. 2014. Los Angeles: Alchemy, 2014. DVD.

Buñuel, Luis, dir. *An Andalusian Dog [Un chien andalou]*. 1929. Paris: Editions Montparnasse, 2005. DVD.

Cameron, James, dir. *The Abyss*. 1989. Los Angeles: 20th Century Studios, 2020. DVD.

———. *Terminator 2: Judgment Day*. 1991. Santa Monica, CA: Lionsgate, 2006. Blu-Ray.

———. *Titanic*. 1997. Hollywood, CA: Paramount, 2014. DVD.

Clark, James B., dir. *Flipper*. 1963. New York: Infobase, 2015. DVD.

Cork, John, dir. *Behind the Scenes with "Thunderball."* 1995. Beverly Hills, CA: MGM Studios, 1995. VHS.

Corman, Roger, dir. *Attack of the Crab Monsters*. 1957. United Kingdom: In2Film, 2010. DVD.

Cousteau, Jacques-Yves, dir. *Sunken Ships* (also translated as *Shipwrecks*) [*Epaves*]. 1943. Poor resolution copy viewable on Vimeo. Accessed December 2020. https://vimeo.com/14157472.

Cousteau, Jacques-Yves, and Louis Malle, dirs. *The Silent World* [*Le monde du silence*]. 1956. United Kingdom: Go Entertainment Group, 2011. DVD.

Cousteau, Philippe, Sr., and Jack Kaufman, dirs. *The Undersea World of Jacques Cousteau*. "Sharks." Aired January 8, 1968, on BBC and ABC. A & E Television Networks, 1991. DVD.

Crisp, Donald, and Buster Keaton, dirs. *The Navigator*. 1924. New York: Kino Video, 1999. DVD.

Del Toro, Guillermo, dir. *The Shape of Water*. 2017. Los Angeles: 20th Century Studios, 2018. DVD.

DeMille, Cecil B., dir. *Reap the Wild Wind*. 1942. New York: Kino Lorber, n.d. Blu-Ray.

Diving to Adventure. Episode 2. With Hans and Lotte Hass. Aired April 13, 1956, on BBC. Archive, Hans Hass Institut für Submarine Forschung und Tauchtechnik.

Diving to Adventure. Episode 3. With Hans and Lotte Hass. Aired April 20, 1956, on BBC. Archive, Hans Hass Institut für Submarine Forschung und Tauchtechnik.

Diving to Adventure. Episode 4. With Hans and Lotte Hass. Aired April 27, 1956. Archive, Hans Hass Institut für Submarine Forschung und Tauchtechnik.

Dunning, George, dir. *Yellow Submarine*. 1968. Los Angeles: Capitol, 2012. Blu-Ray.

Elfick, David, dir. *Crystal Voyager*. 1973. Sydney: Roadshow Entertainment, 2010. DVD.

Fleischer, Richard, dir. *20,000 Leagues under the Seas*. 1954. Burbank, CA: Walt Disney Studios, 2006. DVD.

Friedgan, Raymond, dir. *Killers of the Sea*. 1937. Internet Archive, https://archive.org/details/KillersoftheSea.

Gilbert, Lewis, dir. *The Spy Who Loved Me*. 1977. Beverly Hills, CA: MGM Studios, 2007. DVD.

Gimbel, Peter, and James Lipscomb, dirs. *Blue Water, White Death*. 1971. Santa Monica, CA: Lionsgate, 2013. DVD.

Greenough, George, dir. *The Innermost Limits of Pure Fun*. 1970. The Surf Video Network 2007. DVD.

Gregory, Carl, dir. *Thirty Leagues under the Sea*. With J. E. Williamson. 1914. Submarine Film Corporation-Thanhouser. Current availability unknown.

Hass, Hans, dir. *Abenteuer im Roten Meer*. 1951. Gescher: Polar Film, 2007. DVD.

———. *Menschen unter Haien*. 1945 according to BFI; 1947/48 according to filmportal.de. Gescher: Polar Film, 2007. DVD.

———. *Under the Caribbean* [*Unternehmen Xarifa*]. 1954. British Lion Film Corporation. (English version). Current availability unknown.

———. *Under the Red Sea* [*Abenteuer im Roten Meer*]. 1952. New York: Synergy Entertainment, 2010 (English version). DVD.

———. *Unternehmen Xarifa*. 1954. Gescher: Polar Film, 2007. DVD.

Hitchcock, Alfred, dir. *Psycho*. 1960. Universal City, CA: Universal Pictures, 2020. Blu-Ray.

Kubrick, Stanley, dir. *2001: A Space Odyssey*. 1968. Burbank, CA: Warner Bros., 2018. Blu-Ray.

LeRoy, Mervyn, dir. *Million Dollar Mermaid*. 1952. Burbank, CA: Warner Bros., 2020. Blu-Ray.

Man Ray, dir. *The Mysteries of the Chateau of Dice* [*Les mystères du Château du Dé*]. 1929. Paris: Centre Pompidou, 2007. DVD.

————. *The Starfish* [*L'étoile de mer*]. 1928. Paris: Centre Pompidou, 2007. DVD.

McQuarrie, Christopher, dir. *Mission: Impossible—Rogue Nation*. 2015. Los Angeles: Paramount, 2015. Blu-Ray.

Méliès, Georges, dir. *The Realm of the Fairies* [*Le royaume des fées*]. 1903. Part of *Georges Méliès: First Wizard of Cinema, 1896–1913*. Los Angeles: Flicker Alley, 2008. DVD.

————. *Under the Seas* [*Deux cent milles sous les mers*]. 1907. Part of *Georges Méliès: le premier magicien du cinéma, 1896–1913*. Paris: Lobster Films, 2009. DVD.

————. *Underwater Visit to the* Maine [*Visite sous-marine du Maine*]. 1898. Part of *Georges Méliès: First Wizard of Cinema, 1896–1913*. Los Angeles: Flicker Alley, 2009. DVD.

Milius, John, dir. *Big Wednesday*. 1978. Burbank, CA: Warner Bros., 2018. Blu-Ray. Contains feature-length audio commentary by Milius.

The Monster That Challenged the World. 1957. Trailer. YouTube. Accessed December 2020. https://www.youtube.com/watch?v=h_tOnFhSyXo.

Murnau, F. W. *Nosferatu*. 1922. UK: BFI, 2017. Blu-Ray.

Nichols, Mike, dir. *The Graduate*. 1967. Lawrence Turman Productions. New York: Criterion Collection, 2016. DVD.

Nolan, Christopher, dir. *Inception*. 2010. Burbank, CA: Warner Bros. DVD.

North, Wheeler J., dir. *Rivers of Sand*. 1959. La Jolla, CA: Scientific Diving Consultants. Digitized by the California Audiovisual Preservation Project (CAVPP).

Painlevé, Jean. *The Seahorse* [*L'hippocampe*]. 1934. New York: Criterion Collection, *Science Is Fiction: The Films of Jean Painlevé*, 2009. DVD.

Paton, Stuart, dir. *Twenty Thousand Leagues under the Sea*. 1916. Image Entertainment, 1999. DVD.

Reynolds, Kevin, dir. *Waterworld*. 1995. Pottstown, PA: MVD Entertainment Group, 2019. Blu-Ray.

Rogell, Albert, dir. *Below the Sea*. 1933. New York: Columbia Pictures. Film.

Royal Television Society. "The Making of Blue Planet II." BBC, April 27, 2018. Video. Accessed December 2020. https://www.youtube.com/watch?v=dgwya_Yv_2A.

Schaefer, Armand, dir. *Sixteen Fathoms Deep*. 1934. West Conshohocken, PA: Alpha Video, 2012. DVD.

Sea Hunt. Season 1, episode 1, "Sixty Feet Below." Felix E. Feist, dir. Aired January 4, 1958, in broadcast syndication. Beverly Hills, CA: MGM Studios, 2012. DVD.

————. Season 1, episode 2, "Flooded Mine." Leon Benson, dir. Aired January 11, 1958, in broadcast syndication. Beverly Hills, CA: MGM Studios, 2012. DVD.

————. Season 1, episode 3, "Rapture of the Deep." Anton Leader, dir. Aired January 25, 1958, in broadcast syndication. Beverly Hills, CA: MGM Studios, 2012. DVD.

————. Season 1, episode 4, "Mark of the Octopus." Andrew Marton, dir. Aired February 1, 1958, in broadcast syndication. Beverly Hills, CA: MGM Studios, 2012. DVD.

————. Season 1, episode 5, "The Sea Sled." Lloyd Bridges, dir. Aired February 8, 1958, in broadcast syndication. Beverly Hills, CA: MGM Studios, 2012. DVD

————. Season 1, episode 6, "Female of the Species." Herbert L. Strock, dir. Aired February 15, 1958, in broadcast syndication. Beverly Hills, CA: MGM Studios, 2012. DVD.

————. Season 1, episode 13, "The Shark Cage." Leon Benson, dir. Aired April 5, 1958, in broadcast syndication. Beverly Hills, CA: MGM Studios, 2012. DVD.

————. Season 1, episode 14, "Hard Hat." John Florea, dir. Aired April 12, 1958, in broadcast syndication. Beverly Hills, CA: MGM Studios, 2012. DVD.

———. Season 1, episode 21, "Magnetic Mine." Leon Benson, dir. Aired May 31, 1958, in broadcast syndication. Beverly Hills, CA: MGM Studios, 2012. DVD.

———. Season 1, episode 38, "The Sea Has Ears." Leon Benson, dir. Aired September 27, 1958, in broadcast syndication. Beverly Hills, CA: MGM Studios, 2012. DVD.

———. Season 1, episode 39, "The Manganese Story." Leon Benson, dir. Aired October 4, 1958, in broadcast syndication. Beverly Hills, CA: MGM Studios, 2012. DVD.

———. Season 2, episode 5, "Monte Cristo." Leon Benson, dir. Aired February 1, 1959, in broadcast syndication. Beverly Hills, CA: MGM Studios, 2012. DVD.

———. Season 2, episode 7, "Diving for the Moon." Leon Benson, dir. Aired February 15, 1959, in broadcast syndication. Beverly Hills, CA: MGM Studios, 2012. DVD.

———. Season 3, episode 2, "Water Nymphs." Leon Benson, dir. Aired January 16, 1960, in broadcast syndication. Beverly Hills, CA: MGM Studios, 2012. DVD.

———. Season 3, episode 8, "Missile Watch." Leon Benson, dir. Aired February 27, 1960, in broadcast syndication. Beverly Hills, CA: MGM Studios, 2012. DVD.

———. Season 3, episode 19, "Cross Current." Leon Benson, dir. Aired May 14, 1960, in broadcast syndication. Beverly Hills, CA: MGM Studios, 2012. DVD.

———. Season 4, episode 12, "The Aquanettes." Leon Benson, dir. Aired March 25, 1961, in broadcast syndication. Beverly Hills, CA: MGM Studios, 2012. DVD.

———. Season 4, episode 17, "Niko." Leon Benson, dir. Aired April 29, 1961, in broadcast syndication. Beverly Hills, CA: MGM Studios, 2012. DVD.

Shyamalan, M. Knight, dir. *Lady in the Water*. 2006. Burbank, CA: Warner Bros., 2007. DVD.

Sidney, George, dir. *Jupiter's Darling*. 1955. New York: Infobase, 2016. eVideo.

Spielberg, Steven, dir. *A.I.: Artificial Intelligence*. 2001. Universal City, CA: Dreamworks, 2002. DVD.

———. *Jaws*. 1975. Universal City, CA: Universal Pictures, 2020. DVD.

———. *Minority Report*. 2002. Hollywood, CA: Paramount, 2015. DVD.

———. *Saving Private Ryan*. 1998. Hollywood, CA: Paramount, 2003. DVD.

Sturges, John, and Fred Zinnemann, dirs. *The Old Man and the Sea*. 1958. Burbank, CA: Warner Bros., 1990. VHS.

Sullivan, James R., dir. *Venus of the South Seas*. 1924. Phoenix, AZ: Grapevine Video, 2004. DVD.

Vigo, Jean, dir. *L'Atalante*. 1934. New York: Criterion Collection, 2011. DVD.

———. *Taris, roi de l'eau*. 1931. New York: Criterion Collection, 2011. DVD.

Webb, Robert D., dir. *Beneath the 12-Mile Reef*. 1953. Los Angeles: 20th Century Studios, 2013. DVD.

Wilder, Billy, dir. *Sunset Boulevard*. 1950. Hollywood, CA: Paramount, 2008. DVD.

Yates, Peter, dir. *The Deep*. 1977. Los Angeles: RLJ Entertainment, 2010. Blu-Ray.

Young, Mark, dir. Bonus material in *The Making of 20,000 Leagues under the Sea*. 2003. Burbank, CA: Walt Disney Studios, 2006. DVD.

Young, Terence, dir. *Dr. No*. 1962; Los Angeles, CA: 20th Century Studios, 2012. DVD.

———. *Thunderball*. 1965. Beverley Hills, CA: MGM, 2012. DVD.

Zecca, Ferdinand. *Drama under the Sea* [*Un drame au fond de la mer*]. 1901. Pathé Frères. British Film Institute, Color Fairytales: Early Stencil Films from Pathé, 2012. DVD.

INDEX

A NOTE ON THE TYPE

This book has been composed in Arno, an Old-style serif typeface in the classic Venetian tradition, designed by Robert Slimbach at Adobe.